Econometrics
and Operations Research

V

Herausgegeben von / Edited by

M. Beckmann, Bonn · R. Henn, Göttingen · A. Jaeger, Cincinnati
W. Krelle, Bonn · H. P. Künzi, Zürich
K. Wenke, Ludwigshafen · Ph. Wolfe, Santa Monica (Cal.)

Geschäftsführende Herausgeber / Managing Editors
W. Krelle · H. P. Künzi

The Theory of Max-Min

and its Application to Weapons Allocation Problems

John M. Danskin

With 6 Figures

Springer-Verlag New York Inc. 1967

Dr. John M. Danskin

Center for Naval Analyses of the Franklin-Institute

Arlington, Virginia 22209

Library of Congress Catalog Card Number 66-22462

Titel-Nr. 6480

This book is dedicated to my father
JOHN M. DANSKIN, SR.
who made it possible for
me to become a
mathematician

Preface

Max-Min problems are two-step allocation problems in which one side must make his move knowing that the other side will then learn what the move is and optimally counter. They are fundamental in particular to military weapons-selection problems involving large systems such as Minuteman or Polaris, where the systems in the mix are so large that they cannot be concealed from an opponent. One must then expect the opponent to determine on an optimal mixture of, in the case mentioned above, anti-Minuteman and anti-submarine effort.

The author's first introduction to a problem of Max-Min type occurred at The RAND Corporation about 1951. One side allocates anti-missile defenses to various cities. The other side observes this allocation and then allocates missiles to those cities. If $F(x, y)$ denotes the total residual value of the cities after the attack, with x denoting the defender's strategy and y the attacker's, the problem is then to find

$$\text{Max}_x \text{Min}_y F(x, y) = \text{Max}_x \left[\text{Min}_y F(x, y) \right].$$

If it happens that

$$\text{Max}_x \text{Min}_y F(x, y) = \text{Min}_y \text{Max}_x F(x, y),$$

the problem is a standard game-theory problem with a pure-strategy solution. If however

$$\text{Max}_x \text{Min}_y F(x, y) < \text{Min}_y \text{Max}_x F(x, y),$$

i.e., the order of the choices of x and y is essential, then standard game theory fails. The concept of mixed strategy has no meaning: the x-player knows that his strategy will be observed by his opponent, and the y-player knows x when he acts and needs simply to minimize. Thus the problem needs a separate treatment. That is the object of this book.

It is natural to begin by studying the nature of the function

$$\varphi(x) = \text{Min}_y F(x, y) \tag{*}$$

which is to be maximized. The principal difficulty, illustrated on an example (the "seesaw") in Chapter I, is that $\varphi(x)$ is not in general differentiable in the usual sense, even when $F(x, y)$ is quite smooth. This is closely connected with non-uniqueness in the set $Y(x)$ of values of y

yielding the minimum in (∗). There however is under general conditions on the x- and y-spaces a directional derivative in every direction, given by a formula involving the "answering set" $Y(x)$.* With this result in hand it was then possible to develop a calculus complete with a law of the mean and a Lagrange multiplier theorem for one side condition, but complicated by the lack of a chain rule. Later J. BRAM found a Lagrange multiplier theorem for several side conditions analogous to the well-known Kuhn-Tucker theorem for a simple maximum of a smooth function.

It was now possible to treat a number of problems of Max-Min type. The RAND problem is treated in section 4 of Chapter IV. A problem on the optimal mixture of weapons systems (e. g., Polaris versus Minuteman) was treated for its mathematical interest and then became the theoretical basis of a study on weapons mixtures of the Institute of Naval Studies, the Cambridge, Massachusetts branch of the Center for Naval Analyses. Practical questions led to three-stage problems of Max-Min-Max type. Under some conditions explicit solutions of problems of Max-Min-Max type can be obtained. In the general case information concerning the solution seems to depend on the stability properties of the solutions to the "inside" problems; in the case of "strong forward stability" the solutions become trivial.

By now the material had grown to book-length. The writing was supported at the Center for Naval Analyses, to whose Director, Dr. FRANK BOTHWELL, I am grateful for the stimulus to write the book and for the time to do it. I wish to express also my thanks to Dr. JOSEPH BRAM and Mrs. SIDNIE FEIT, who read and criticized previous versions of portions of the book; Dr. BRAM has been kind enough to allow me the use of the material in the Appendix. There is finally Mrs. JANE LYNDE of the Institute of Naval Studies in Cambridge, Massachusetts, who typed the many versions of this book. The author is also very grateful to Captain DANIEL J. MURPHY, U.S.N. of the Office of the Chief of Naval Operations, whose interest and support greatly facilitated the writing of this book.

JOHN M. DANSKIN

---

* After this book was finished, the author's attention was called to a somewhat similar theorem on the derivative of the value of a game, due to OLIVER GROSS and referred to in a paper [7] by HARLAN MILLS. The Gross theorem referred to matrix games and continuous games over the square. See also the addendum to [8].

Table of Contents

CHAPTER I

Introduction

This book is concerned with a problem arising frequently in practice. There are two antagonists. One must act first, knowing that the second will learn what he has done and then act to *his* best advantage. What should the first do?

For instance, suppose one antagonist must select a strategic weapons mixture consisting of various numbers of various types of weapons. If the antagonist is the United States, he considers, for example, selecting a mixture consisting of so many Polaris submarines with a certain weapon, so many Minutemen missiles with a certain warhead and a certain silo design, and so many of some other type of weapon.

Whoever the antagonist in this example might be, he must select his weapons mixture knowing that his opponent will learn what it is and will try to develop a suitable countering mixture. Thus, in the example above, an opponent might purchase so much anti-submarine equipment to counter the Polaris, so many missiles to strike the Minutemen in their silos, and so forth.

It is easy to say, if the first "player" is the United States, how an opponent might learn what his weapons mixture is. The Defense Appropriations Hearings of the House and Senate are detailed in both general characteristics and costs. The trade journals of the defense industries give specific technical details and production schedules. Strategic quarrels are leaked to newspapers and discussed in political campaigns.

One should not draw from the above remarks the conclusion that the author feels that the public disclosures made in the United States are necessarily bad for it. If they were not made, information would leak through contracts, many, like those for submarine construction or silos, vast in magnitude and involving multitudes of workers.

A country with a policy of complete secrecy must also expect that its opponent will learn its mixture, for the same reasons. No really large program will fail to involve disaffected or self-important workers who in one way or another, directly or indirectly, make this knowledge known to the opponent. Nor can really large structures be hidden entirely from the prying eye of the satellites and other optical devices, nor large nuclear detonations be made without giving sufficient debris for the whole world to know very closely how the bomb was made.

Thus in the really large-scale matters such as decisions as to weapons mixes, *any* country must assume that its opponent will find out what it has done and counter.

At this stage it must be decided what is meant by a counter. And before this is possible it must be decided what the object of the first "player" is.

In this book it will be supposed that the first player has a single definite object, and that the second player directly opposes this object. Thus, the first player in the example of Chapter V wishes to maximize the damage to his opponent, the second to minimize this damage.

Except in Chapters VI and VII, we are concerned with two-move situations. Player 1 moves, wishing to maximize, and then player 2 moves, wishing to minimize.

Possibly the most important measure of the effectiveness of a strategic weapons system is its retaliatory capability. Suppose, for instance, that a weapons mixture consists originally of x_1 weapons of type 1, x_2 weapons of type 2, and finally x_n weapons of type n. We shall write

$$x = (x_1, \ldots, x_n)$$

to denote this mixture. An opponent, on learning this mixture, purchases a counter[1]

$$y = (y_1, \ldots, y_n)$$

with the quantity y_i opposing the x_i. The counter is then applied to the mixture x, yielding a residual mixture $x' = (x'_1, \ldots, x'_n)$. This is the mixture left for retaliation.

It might be possible to assign a definite value v_i to each unit of the residual weapons of type i. Then the value of the residual weapons system would be

$$V(x, y) = \Sigma v_i x'_i.$$

The object of the x-player, who acts first by making his purchase, is to maximize the total residual value $V(x, y)$. The object of the y-player, who acts second and in full knowledge of x, is to minimize $V(x, y)$. Thus, if the first player makes a choice x from the strategies open to him, he must assume that the second player will act to minimize and that the result will be

$$\varphi(x) = \operatorname*{Min}_{y} V(x, y).$$

Here the minimum is taken over the space of strategies open to the second player. The object of the x-player is to maximize $\varphi(x)$. He thus

---

[1] One counter-weapon may oppose more than one weapons type; for instance, anti-submarine warfare may oppose several types of submarine systems. We shall treat an example of this in Chapter V, under the heading "simultaneously vulnerable systems."

seeks the x yielding

$$\text{Max}_{x} \; \varphi(x) = \text{Max}_{x} \; \text{Min}_{y} \; V(x, y). \tag{1}$$

This is the reason for the title of this book. It deals with problems arising from successive choices by two diametrically opposed "players."

The book does not, it should be said at once, deal with all such problems. $V(x, y)$ is continuous (and more) in all our discussions. Thus we do not attempt to treat successive-move common games such as chess with win-or-lose outcomes. Generally speaking, while some of the results given in this book are stated more generally, the x and y are real-valued vectors in Euclidean space, representing allocations of effort, and the result of the allocations varies continuously with them.

The two-move problem (1) is by no means adequate to describe the true problem of weapons selection for retaliation. It supposes that a value can be placed on a unit element of a weapons mixture. This is to some extent possible when there are few weapons and many targets, so that each weapon may be valued by the damage it does to the best target available to it. But when there are more weapons than targets, the situation becomes more complex. How to measure the value of the third of five weapons of different kinds arriving at a target?

In fact, as soon as nonlinear effects enter ("overkill" or "saturation") one cannot find a single clear measure of the value of an individual weapon appearing in a mix. Generally speaking, one can only value the mix itself, and that by the damage it does to a given target system. This damage in turn must be associated with a particular allocation of the weapons to the target system. It is necessary to choose such a particular allocation; we shall choose it to be the optimal, or maximizing allocation.

Thus, instead of problem (1) above we have the following situation: the first player picks $x = (x_i, \ldots, x_n)$ from the strategies available to him. The second picks $y = (y_1, \ldots, y_n)$, and applies it to x, yielding $x' = (x'_i, \ldots, x'_n)$. This is the initial strike of a war. The x-player is assumed to know the residuals $x'_1, \ldots, x'_n$. He then allocates these to targets by a matrix

$$\xi = \|\xi_{ik}\|, \quad i = 1, \ldots, n; \; k = 1, \ldots, K. \tag{2}$$

Here ξ_{ik} denotes the proportion of the residual quantity x'_i applied to the kth target, so that

$$\Sigma_k \xi_{ik} = 1, \quad i = 1, \ldots, n. \tag{3}$$

The damage to the target system may then be described in terms of a function $D(x, y, \xi)$ of three variables corresponding to the successive actions.[2]

---

[2] The particular form used for $D(x, y, \xi)$ is discussed in Chapter VII.

1*

We formulate the problem "from inside out." The first player, wishing to maximize, chooses x knowing that there are two more steps. The second player then chooses y knowing that the residuals x_i' will be optimally applied. The first player, observing the residuals, then maximizes with respect to the allocation ξ. This last step is finding the ξ yielding

$$H(x, y) = \underset{\xi}{\text{Max}}\ D(x, y, \xi),$$

given x and y. The next-to-last step is finding the y yielding

$$G(x) = \underset{y}{\text{Min}}\ H(x, y) = \underset{y}{\text{Min}}\ \underset{\xi}{\text{Max}}\ D(x, y, \xi),$$

given x.

The first step is finding the x yielding

$$\underset{x}{\text{Max}}\ G(x) = \underset{x}{\text{Max}}\ \underset{y}{\text{Min}}\ \underset{\xi}{\text{Max}}\ D(x, y, \xi). \tag{4}$$

Thus we arrive at a Max-Min-Max problem, corresponding to a "simplest possible" approach to a realistic model for the optimal choice of a weapons mix for strategic retaliation.

A technical point arises at this stage of the discussion, which must be noted here. Throughout the book we deal with real numbers rather than integers. Our main purpose will be to find *principles* rather than numerical allocations, though some problems are solved explicitly. In any applied problem the allocation matrix must be checked for common sense, as for instance if an allocation were to send .002 one-thousand-megaton bombs to some target. It is necessary, as we have said above, to *consider* the allocation mix ξ to give a value to the weapons mix. It is not necessary to *make* the allocation at the epoch when the mix $x = (x_1, \ldots, x_n)$ is being chosen. It is indeed the latter which is of prime interest to us.

Let us return from this apology to consider another objective for the first player: to limit city damage on his own country, by striking the other side's weapons as the first move of a war. Such an action may well seem justifiable to a country which has just learned at 12 noon with certainty that the opponent is going to strike at 2 p.m., and finds communication channels out or, using them, meets only with hostility and taunts. Suppose then the original purchase $x = (x_1, \ldots, x_n)$ is made with this possibility (and the following scenario) as dominant. The second player is supposed as always diametrically opposed; he observes x and chooses city-busting weapons $y = (y_1, \ldots, y_m)$ knowing that the mixture x will be allocated to those weapons. The first player now carries out his preemptive damage-limiting strike by striking at the weapons mixture y only. This is done by choosing a matrix ξ as above. The second player now observes his residuals $y' = (y_1', \ldots, y_m')$, which are functions of ξ, and allocates these to the cities of the first player by an allocation matrix η

of the same general form as ξ. We thus arrive at the problem

$$\text{Min}_{x} \text{Max}_{y} \text{Min}_{\xi} \text{Max}_{\eta} D(x, y, \xi, \eta) \tag{5}$$

which represents a simplest-possible model for weapons selection for pre-emptive damage-limiting.

Thus we begin to distinguish hierarchies of models. First there is the simplest possible model. If the objective of the first player on purchasing the mixture $x = (x_1, \ldots, x_n)$ is a city-busting first strike on his opponent, with no other consideration, then the counter will never be applied and we have a pure maximum problem:

$$\text{Max}_{x, \xi} D(x, \xi). \tag{6}$$

This model is equivalent to the allocation model for the matrix ξ alone. Then there is the linear retaliation model (1):

$$\text{Max}_{x} \text{Min}_{y} V(x, y).$$

and then the non-linear retaliation model (4):

$$\text{Max}_{x} \text{Min}_{y} \text{Max}_{\xi} D(x, y, \xi),$$

and the pre-emptive damage-limiting model (5):

$$\text{Min}_{x} \text{Max}_{y} \text{Min}_{\xi} \text{Max}_{\eta} D(x, y, ,\xi, \eta).$$

The retaliation models will naturally, all other things being equal, favor the less vulnerable weapons. Models (5) and (6) may well favor quite vulnerable weapons.

One may, of course, propose such "scenarios" of any length, all comprised in a 2m-step model

$$\text{Max}_{x^1} \text{Min}_{y^1} \ldots \text{Max}_{x^m} \text{Min}_{y^m} F(x^1, y^1, \ldots, x^m, y^m) \tag{7}$$

with the vector x^j in some Euclidean space R^{n_j} and y^j in some Euclidean space S^{p_j}, with in general no special pattern relating the various R^{n_j} and S^{p_j}, $j = 1, \ldots, m$.

The object of this book is to treat the Max-Min problem in a systematic fashion, and to make some steps toward more complex problems, such as the Max-Min-Max problem of Chapter VII.

The basic technical difficulty arises already when there are only two steps. This is that even if the function $F(x, y)$ of the two variables x, y is quite smooth, the derivative of the function

$$\varphi(x) = \text{Min}_{y} F(x, y)$$

may well not exist. A very easy example is the "seesaw" example

$$F(x, y) = y \sin x \tag{8}$$

with $-\dfrac{\pi}{2} \le x \le \dfrac{\pi}{2}$ and $-1 \le y \le +1$.

The x-player chooses the angle (see Fig. 1) of the seesaw, the y-player a point somewhere between the left end $y = -1$ and the right end $y = +1$. The function $F(x, y)$ measures the height of the point chosen by the y-player above the fulcrum. Obviously

$$\varphi(x) = -|\sin x| \tag{9}$$

and the function φ is not differentiable at the point $x = 0$, which is the solution of the Max-Min problem.

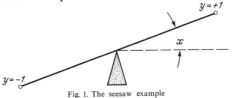

Fig. 1. The seesaw example

This difficulty is overcome by the introduction of a *directional* derivative, which is the principal tool of the theory. Suppose $F(x, y)$ is a function of the vector x in a Euclidean space R^n and the element y in a compact topological space $\mathscr{Y}$. Suppose $F(x, y)$ and $F_{x_i}(x, y)$ are continuous in $R^n \times \mathscr{Y}$ for all i. Finally let $\gamma = (\gamma_1, \ldots, \gamma_n)$, with $\gamma_1^2 + \cdots + \gamma_n^2 = 1$, be any direction in R^n. Put

$$\varphi(x) = \underset{y}{\text{Min}}\, F(x, y). \tag{10}$$

Then the directional derivative $D_\gamma \varphi(x)$ of φ at the point x exists and is given by

$$D_\gamma \varphi(x) = \underset{y \in Y(x)}{\text{Min}} \sum_{i=1}^{n} \gamma_i F_{x_i}(x, y), \tag{11}$$

where $Y(x)$ denotes the set of y yielding the minimum against the given x in (10).

For instance, in the trivial problem (8), let us use (11) to calculate the right derivative of $\varphi(x)$ at $x = 0$. We have $\gamma = 1$ and $F_x(x, y) = y \cos x$, so that from (11)

$$D_R \varphi(x) = \underset{y \in Y(0)}{\text{Min}}\, y \cos 0 = \underset{-1 \le y \le +1}{\text{Min}}\, y = -1,$$

as is seen also from (9). In checking the left derivative, the reader should recall that this is the negative of the directional derivative in the direction -1.

This directional derivative lacks several properties of the ordinary derivative. In particular it does not obey the chain-rule of elementary differentiation. In Chapter III we give an example of a smooth function of four real variables $F(x_1, x_2, y_1, y_2)$ for which the corresponding function $\varphi(x_1, x_2) = \underset{y_1, y_2}{\text{Min}}\, F(x_1, x_2, y_1, y_2)$ has at certain points a zero derivative in the $+x_1$ direction, a zero derivative in the $+x_2$ direction, and a positive derivative in the 45° direction. This lack of a chain-rule is a striking feature of the Max-Min theory: *in solving a Max-Min problem one cannot in general reduce the problem to differentiation in the directions of the coordinate axes.*

The usual proofs of the Lagrange multiplier principle for the simple maximum problem make explicit use of the derivatives in the directions of the axes. It is therefore possible to wonder whether there might be no general Lagrange multiplier principle for problems of Max-Min type. In fact there is a quite general Lagrange multiplier principle, as general for Max-Min problems as the so-called Kuhn-Tucker principle is for Max problems. The simplest version of a Lagrange multiplier principle is for the conditions

$$\Sigma x_i = X, \; x_i \geq 0. \tag{12}$$

It is worthwhile quoting the result here. If $\varphi(x)$ is maximized at $x = x^0$ under the side conditions (12), then there exists a λ such that

$$D_\gamma[\varphi(x) - \lambda \Sigma x_i] \leq 0 \tag{13}$$

for any possible (i.e. $\gamma_i \geq 0$ if $x_i^0 = 0$) direction whatever.

The proof in the above case is extremely simple and makes use of the superadditivity principle mentioned below and of the Dedekind cut property of the real numbers: it is due to the author (Chapter III). The proof of the result with several non-linear side conditions is quite different and is due to JOSEPH BRAM (Appendix).

The superadditivity principle is the following. Suppose α, β, and γ are unit vectors with

$$\gamma = a\alpha + b\beta, \; a, \; b \geq 0.$$

Then

$$D_\gamma \varphi(x) \geq a D_\alpha \varphi(x) + b D_\beta \varphi(x). \tag{14}$$

This result, which follows immediately from formula (11), is the replacement for the chain rule. That it is sometimes a strict inequality is a consequence of the four-variables example mentioned above, in which the left side is positive and the right side zero.

There is also a replacement for the law of the mean, which is proved using an essentially trivial result on one-sided derivatives of functions of one variable, which the author has not seen in the literature.

Chapter II gives some elementary maximum theory and treats some examples of pure-strategy-solution games. These examples, an extension of notes prepared by the author in 1950–3 and hitherto unpublished, are not, by now, new. There are two purposes in presenting them. The first is to explain what kinds of solutions are wanted in the very much more difficult Max-Min, Max-Min-Max, etc., problems. The second is to present facts specifically needed in Chapter V. This chapter may be read by anyone with a firm grasp of the ordinary derivative.

Chapter III presents the basic facts, mostly mentioned above, concerning the directional derivative.

Chapter IV gives four examples of Max-Min problems. The first treats formally the four-variable example mentioned above. The second is one of the games of Chapter II treated by the directional derivative technique. The third is an honest Max-Min Problem (i.e., sometimes Min Max > Max Min) and has not been treated before. The solution to the "inside" Min problem is not unique and exhibits unexpected stability properties. The fourth problem found the author's introduction to the subject; it was posed at the Rand Corporation around 1951. The result presented is the first theoretical result to be obtained for this problem. The Lagrange multiplier principle is of little use in this problem, and, the chain rule not holding, it was the author's first explicit use of differentiation off the directions of the axes.

Chapter V presents a solution of a problem posed around 1960 by THOMAS E. PHIPPS with the objective of treating differences, previously considered "qualitative", between the sea-based systems and land-based systems, in a quantitative way. The solution consists in theorems giving its general structure, and a mathematically exact procedure for obtaining numerical solutions in any given case. This model has played an important rôle in discussions of various military systems. Its importance in the applications is enhanced by the extraordinary fact noted in Chapter VII: the most general Max-Min-Max models, otherwise only partially solvable, reduce to this case when their coefficients decompose in a certain way.

Chapter VI treats the allocation problem for the matrix ξ mentioned above (2). This material is essential for the further development; in addition it is of independent value and applies to many practical situations. The method of obtaining the solution is described in detail sufficient for a programmer to make a program, and the structure of the solution treated in a number of lemmas. Some examples are worked.

Chapter VII treats the Max-Min-Max problem. For the general problem only partial results are obtained. In one important case the problem reduces to the Phipps model of Chapter V, so that a non-linear problem reduces to a linear one. This is the case of decomposable coefficients, and has at present very wide application to real problems, since many bomb-damage formulas in common use have decomposable coefficients.

In the general case, when only "percentage vulnerable" (i.e., $x'_i = x_i e^{-\alpha_i y_i}$) systems are being considered, there is a sufficiency test for admission to the mix, involving the effectiveness and the vulnerability.

The question as to the quantity of the mix in the optimal system is more difficult. In some cases it is possible to say something, as in the case of forward stability or weak forward stability. In these cases the weapons systems which appear in the optimal mix and which are attacked appear in amounts inversely proportional to their vulnerability. But forward stability is not a general phenomenon, as we show by example. The whole question of stability is evidently central to this and to other problems of Max-Min type, and constitutes in the author's view the deepest part of the theory, and at the same time the least developed.

The appendix is a proof of the Lagrange multiplier principle for Max-Min with several side conditions, due to JOSEPH BRAM. This generalizes the result first obtained by the author for one linear side condition.

CHAPTER II

Finite allocation games

1. Objective

We have stated that we are mainly after principles rather than computational methods. The principles are sometimes stated in terms of theorems, sometimes in terms of explicit formulas. We use real variable theory in its exact rigor. Thus, while we must in any application admit approximation in the formulas we start with, we do not admit any in our analyses. This allows to build mathematical structures of more than one stage, as, for instance, in Chapter VII, and to go rather further than one might expect without resorting to machine methods.

This chapter has a triple object. It forms an elementary introduction to finite allocation games; using these games as examples it explains what kinds of principles are sought in the harder problems; and it presents specific results needed in Chapter V.

The chapter contains one game fully worked out, and another in the exercises.

2. The Gibbs Lemma

Before passing to allocation games we deal with the simplest maximum problem and some examples.

We need a simple lemma which will be called the Gibbs lemma, since it was essentially known to J. WILLARD GIBBS.[3]

Gibbs lemma. *Suppose the set* $x^0 = (x_1^0, \ldots, x_n^0)$ *maximizes* $\Sigma_i f_i(x_i)$ *subject to the side conditions* $\Sigma x_i = X, x_i \geq 0$. *Suppose the* f_i *differentiable. Then there exists a* λ *such that*

$$f_i'(x_i^0) = \lambda \ \ if \ x_i^0 > 0$$
$$\leq \lambda \ \ if \ x_i^0 = 0. \tag{1}$$

Proof. Suppose $x_i^0 > 0$ and $j \neq i$. Let $0 \leq \varepsilon < x_i^0$ and put

$$\varphi(\varepsilon) = f_i(x_i^0 - \varepsilon) + f_j(x_j^0 + \varepsilon) + \sum_{k \neq i, j} f_k(x_k^0). \tag{2}$$

Evidently the set x^0 so altered still satisfies the side condition. $\varphi(\varepsilon)$ is a differentiable function of ε. As $\varphi(\varepsilon)$ is a maximum at $\varepsilon = 0$, evidently

---

[3] The 1876 paper on "Equilibrium of Heterogeneous Substances", in the *Collected Works*, Yale University Press.

$\varphi'(0) \leqq 0$. On referring to (2) this yields

$$f_j'(x_j^0) \leqq f_i'(x_i^0). \tag{3}$$

But the only hypothesis was that $x_i^0 > 0$. Thus if both $x_i^0 > 0$ and $x_j^0 > 0$ the reverse of (3) holds as well. Hence all those $f_i'(x_i^0)$ with $x_i^0 > 0$ have a common value λ, and, from (2), $f_i'(x_i^0) \leqq \lambda$ if $x_i^0 = 0$.

The lemma is proved.

This lemma (along with its generalizations) might properly be called the fundamental lemma of mathematical operations research. It is also central to normative economics — it is the marginal utility principle, and is the theoretical basis for price formation. It is analogous to the Euler equations in the classical calculus of variations.

3. A maximizing example

We shall use the Gibbs lemma to solve a simple maximum problem:

$$\text{Maximize } \Sigma a_i(1 - e^{-b_i x_i}) \tag{4}$$

subject to

$$\Sigma x_i = X, \; x_i \geqq 0. \tag{5}$$

We suppose all the a_i and b_i positive. Writing $f_i(x_i) = a_i(1 - e^{-b_i x_i})$, we find from (1) that the solution satisfies, for some λ,

$$\begin{aligned} a_i b_i e^{-b_i x_i} &= \lambda \quad \text{if} \quad x_i > 0; \\ &\leqq \lambda \quad \text{if} \quad x_i = 0. \end{aligned} \tag{6}$$

We drop the superscript denoting the solution in all our discussions when this will not cause confusion. The key to this problem is the relative value of $a_i b_i$ and λ.

If $a_i b_i \leqq \lambda$ we could not have $x_i > 0$, for, if we did, the top line must hold in (6); but it cannot hold since $e^{-b_i x_i} < 1$. Hence if $a_i b_i \leqq \lambda$, $x_i = 0$.

Suppose $a_i b_i > \lambda$. Then $x_i = 0$ is impossible, since the bottom line of (6) then asserts that $a_i b_i \leqq \lambda$. It follows that $x_i > 0$ and the top line holds in (6). But this equation is trivially solved:

$$x_i = \frac{1}{b_i} \log \frac{a_i b_i}{\lambda} \quad \text{if} \quad a_i b_i > \lambda. \tag{7}$$

We have thus solved the problem (4), except for the determination of λ. This is gotten by substituting into the side condition (5): λ must satisfy

$$\sum_{a_i b_i > \lambda} \frac{1}{b_i} \log \frac{a_i b_i}{\lambda} = X. \tag{8}$$

Equation (8) is quite simple to solve. Denote its left side by $h(\lambda)$. Put $\mu = \text{Max}_i \, a_i b_i$, $v = \text{Min}_i \, a_i b_i$. If $\lambda < v$ all the entries $i = 1, ..., n$ are present in the sum on that left side, and if λ is sufficiently small $h(\lambda)$ exceeds any given positive constant, in particular X. If on the other hand $\lambda \geqq \mu$, we have $h(\lambda)=0$. It is evident that $h(\lambda)$, for $0<\lambda \leqq \mu$, is strictly decreasing, since the terms decrease individually and their number changes only by dropping when λ passes through a value of $a_i b_i$. Finally, $h(\lambda)$ is continuous, since when λ passes through a value of $a_i b_i$ the terms which are dropped have all just reached zero.

The function $h(\lambda)$ thus must pass through the value X exactly once. The resulting λ is the desired one. In practice just this procedure is very easily accomplished, so that for this problem the solution is completely explicit.

This example affords an opportunity to explain the meaning of "principles." Suppose there were a number of targets of various degrees of vulnerability, and of various values. We suppose that the proportionate damage done the ith target with a level x_i of attack is $1-e^{-b_i x_i}$, and that the value of that target is a_i. The numbers b_i will be called the *vulnerabilities*. The total damage done on all the targets in the complex is thus given by the sum in (4), on the assumption that damage is additive. Suppose the total amount of effort available is X. Then the question is, how best to distribute that effort? I.e., which targets should one attack, and at what level? This is of course the problem we have just solved. The answer may be put into words. We attack only the targets for which the product of the value a_i and the vulnerability b_i is larger than some quantity λ. *The criterion for attacking or not attacking is* $a_i b_i$. The exact critical levels are to be found from formula (8).

In the above instance we have an explicit solution and the "principles" taken alone are therefore not very interesting. But as our problems get harder explicit solutions are harder to find and the principles correspondingly more interesting. Sometimes they are striking, as principle (ii) in the game worked out in Section 4. Even in the above example it was not *a priori* obvious that the particular combination $a_i b_i$ was the criterion.

The above example is the finite analog of the Koopman search problem solved in his paper [5], page 617. The analogy is furnished by the following interpretation: an object is in one of a number of boxes with probabilities $a_i, ..., a_n$. If the searcher spends x_i time units looking for it in the ith box he will find it with probability $1-e^{-b_i x_i}$. Then (4) represents the probability he will find it if he spends the times $x_1, ..., x_n$ in the various boxes. If his total time available is X, then the problem is the one done above. See Exercise 1.

4. A finite allocation game

In this section we shall study in full detail a finite allocation game of importance throughout this book.

Put

$$F(x, y) = \sum_{i=1}^{n} v_i x_i e^{-\alpha_i y_i}, \tag{9}$$

where $x = (x_1, \ldots, x_n)$ and $y = (y_1, \ldots, y_n)$, and all the v_i's and α_i's are positive. The x-player seeks to maximize $F(x, y)$ and the y-player to minimize; we suppose these players to be subjected respectively to the conditions

$$x_1 + \cdots + x_n = X, \quad x_i \geqq 0 \tag{10}$$

and

$$y_1 + \cdots + y_n = Y, \quad y_i \geqq 0, \tag{11}$$

with $X > 0$ and $Y > 0$.

The function $F(x, y)$ given by (9) is linear and therefore concave in the maximizing player, and convex in the minimizing player. It accordingly[4] has a solution, i.e. a pair of optimal strategies x^0, y^0 such that

$$F(x^0, y) \geqq F(x^0, y^0) \quad \text{for all} \quad y \tag{12}$$

and

$$F(x, y^0) \leqq F(x^0, y^0) \quad \text{for all} \quad x. \tag{13}$$

We may restate (12) as follows: y^0 minimizes the expression

$$\Sigma v_i x_i^0 e^{-\alpha_i y_i}$$

subject to (11). From the Gibbs lemma (1) (and making a change of sign) we find that there exists a μ such that

$$v_i \alpha_i x_i^0 e^{-\alpha_i y_i^0} = \mu \quad \text{if} \quad y_i^0 > 0;$$
$$\leqq \mu \quad \text{if} \quad y_i^0 = 0. \tag{14}$$

Similarly we find from (13) that there exists a λ such that

$$v_i e^{-\alpha_i y_i^0} = \lambda \quad \text{if} \quad x_i^0 > 0;$$
$$\leqq \lambda \quad \text{if} \quad x_i^0 = 0. \tag{15}$$

We carry out the solution by means of a series of lemmas.

Lemma 1. $\mu > 0$ and $\lambda > 0$.

---

[4] The fundamental theorem of the theory of games in the form appropriate here is the following. *Suppose $F(x, y)$ is a continuous function on a space $K \times L$, where K and L are compact and convex. Suppose that the set of points $X(y)$ yielding the maximum to $F(x, y)$ for y fixed is convex for each such y and that the set of points $Y(x)$ yielding the minimum to $F(x, y)$ for x fixed is convex for each such x. Then there is a pure-strategy solution of the form* (12)–(13). The theorem in this form is proved in KAKUTANI [4]. Its application to the case at hand is the following: if $F(x, y)$ is concave in x and convex in y, the sets $X(y)$ and $Y(x)$ are indeed convex (Exercise 3).

Proof. Since $X > 0$, some $x_i^0 > 0$. The left side of (14) for that value of i is therefore positive, and hence $\mu > 0$. That $\lambda > 0$ is obvious.

Lemma 2. $y_i^0 > 0$ *implies* $x_i^0 > 0$.

Proof. If $y_i^0 > 0$, the top line of (14) holds, so that, since $\mu > 0$, $x_i^0 > 0$.

Lemma 3. *If* $v_i < \lambda$, $x_i^0 = y_i^0 = 0$.

Proof. It is only necessary, from Lemma 2, to prove $x_i^0 = 0$. If $x_i^0 > 0$ the top line of (15) holds, which is impossible since $v_i e^{-\alpha_i y_i^0} \leq v_i < \lambda$. Hence $x_i^0 = 0$.

Lemma 4. *If* $v_i = \lambda$, $y_i^0 = 0$ *and* $x_i^0 \leq \dfrac{\mu}{\alpha_i \lambda}$.

Proof. Suppose $y_i^0 > 0$. Then $x_i^0 > 0$ from Lemma 2, and equality holds in (15). But this is impossible since $v_i e^{-\alpha_i y_i^0} < v_i = \lambda$. Hence $y_i^0 = 0$ as asserted. The second statement follows from the bottom line of (14).

Lemma 5. *If* $v_i > \lambda$, *both* x_i^0 *and* y_i^0 *are positive and*

$$x_i^0 = \frac{\mu}{\alpha_i \lambda} ; \tag{16}$$

$$y_i^0 = \frac{1}{\alpha_i} \log \frac{v_i}{\lambda} . \tag{17}$$

Proof. Suppose $y_i^0 = 0$. Then $v_i e^{-\alpha_i y_i^0} = v_i > \lambda$, which contradicts (15). Hence $y_i^0 > 0$, and, from Lemma 2, $x_i^0 > 0$. We now have equality in both (14) and (15). (16) results from dividing (15) into (14). (17) is the solution of (15) for y_i^0.

Lemma 6. λ *is unique and satisfies the equation*

$$\sum_{v_i > \lambda} \frac{1}{\alpha_i} \log \frac{v_i}{\lambda} = Y . \tag{18}$$

Proof. From Lemmas 3 and 4, $y_i^0 > 0$ if and only if $v_i > \lambda$; and in that case, from Lemma 5, y_i^0 is given by (17). The result follows.

Lemma 7. μ *satisfies the inequality*

$$\frac{\lambda X}{\displaystyle\sum_{v_i \geq \lambda} \left(\frac{1}{\alpha_i} \right)} \leq \mu \leq \frac{\lambda X}{\displaystyle\sum_{v_i > \lambda} \left(\frac{1}{\alpha_i} \right)} . \tag{19}$$

Proof. From Lemmas 3 and 5,

$$\sum_{v_i > \lambda} \frac{\mu}{\lambda \alpha_i} + \sum_{v_i = \lambda} x_i^0 = X . \tag{20}$$

Since, from Lemma 4, $0 \leq x_i^0 \leq \dfrac{\mu}{\lambda \alpha_i}$ when $v_i = \lambda$, it follows from (20) that

$$\sum_{v_i > \lambda} \frac{\mu}{\lambda \alpha_i} \leq X \leq \sum_{v_i \geq \lambda} \frac{\mu}{\lambda \alpha_i} , \tag{21}$$

which is equivalent to (19).

Lemma 8. *The value of the game is* λX.
Proof. From (15),

$$F(x^0, y^0) = \sum_{i=1}^{n} v_i x_i^0 e^{-\alpha_i y_i^0} = \sum_{x_i^0 > 0} v_i x_i^0 e^{-\alpha_i y_i^0} = \lambda \sum_{x_i^0 > 0} x_i^0 = \lambda X . \quad (22)$$

Denote by μ_0 and μ_1 the left and right extremes in (19) respectively. The following lemma describes the ambiguity in the solution for x.

Lemma 9. *Let* $\mu_0 \leq \mu^* \leq \mu_1$, *so that* (21) *holds. For* $v_i > \lambda$ *put* $x_i^* = \dfrac{\mu^*}{\lambda \alpha_i}$.

For $v_i = \lambda$ *choose* x_i^* *with* $0 \leq x_i^* \leq \dfrac{\mu^*}{\lambda \alpha_i}$ *any way so that*

$$\sum_{v_i > \lambda} \frac{\mu^*}{\lambda \alpha_i} + \sum_{v_i = \lambda} x_i^* = X , \quad (23)$$

which is possible because of (21). *For* $v_i < \lambda$ *put* $x_i^* = 0$. *Then the vector* $x^* = (x_1^*, \ldots, x_n^*)$ *is optimal for the maximizing player.*

Proof. We need to prove that $F(x^*, y) \geq \lambda X$ for any y satisfying (11). As in (22), $F(x^*, y^0) = \lambda X$. Let y^1 satisfy (11). Put $y^t = (1-t)y^0 + ty^1$, with $0 \leq t \leq 1$. Define

$$\varphi(t) = F(x^*, y^t) = \sum_{v_i \geq \lambda} v_i x_i^* e^{-\alpha_i y_i^t} .$$

We have

$$\varphi'(t) = \sum_{v_i \geq \lambda} \alpha_i v_i x_i^* e^{-\alpha_i y_i^t} \cdot (y_i^0 - y_i^1) \quad (24)$$

and

$$\varphi''(t) = \sum_{v_i \geq \lambda} \alpha_i^2 v_i x_i^* e^{-\alpha_i y_i^t}(y_i^0 - y_i^1)^2 \geq 0 . \quad (25)$$

From (24),

$$\varphi'(0) = \sum_{v_i \geq \lambda} \alpha_i v_i x_i^* e^{-\alpha_i y_i^0}(y_i^0 - y_i^1)$$

$$= \lambda \sum_{v_i \geq \lambda} \alpha_i x_i^*(y_i^0 - y_i^1)$$

$$= \sum_{v_i > \lambda} \mu(y_i^0 - y_i^1) + \lambda \sum_{v_i = \lambda} \alpha_i x_i^*(y_i^0 - y_i^1)$$

$$= \mu \sum_{v_i \geq \lambda} (y_i^0 - y_i^1) + \lambda \sum_{v_i = \lambda} (\alpha_i x_i^* - \mu/\lambda)(y_i^0 - y_i^1) . \quad (26)$$

But if $v_i = \lambda$ we have $y_i^0 = 0$ and $\alpha_i x_i^* - \mu/\lambda \leq 0$, so that the right term in the last expression in non-negative. Hence the calculation (26) may be continued to

$$\varphi'(0) \geq \mu \sum_{v_i \geq \lambda} (y_i^0 - y_i^1) = \mu [Y - \sum_{v_i \geq \lambda} y_i^1] \geq 0 . \quad (27)$$

Because of (25), we have from (27) that $\varphi'(t) \geqq 0$ for all $t \in [0, 1]$. Hence $\varphi(1) \geqq \varphi(0)$, i.e., $F(x^*, y^1) \geqq F(x^*, y^0) = \lambda X$. The lemma is proved.

The solution of (9) is thus complete. One first determines λ from (18). The interval (μ_0, μ_1) is then given by (19). One takes any μ on this interval and writes $x_i^0 = \dfrac{\mu}{\lambda \alpha_i}$ if $v_i > \lambda$, and then choosing x_i^0 in the case $v_i = \lambda$ so that $x_i^0 \leqq \dfrac{\mu}{\lambda \alpha_i}$ and so that (20) is satisfied. This is a solution for the maximizing player; the solution for the minimizing player is given uniquely by $y_i^0 = \dfrac{1}{\alpha_1} \log \dfrac{v_i}{\lambda}$ for $v_i > \lambda$, $y_i^0 = 0$ otherwise.

Stability. Let us discuss the stability of (9) with respect to variations in X and Y. As to variations in X alone, these do not affect the solution for y at all. The two ends of the interval (19) of possible μ vary continuously in X. Thus the family of solutions for x described by Lemma 9 varies continuously in X in the sense that its bounds vary continuously. With respect to Y the situation is different. If λ is not equal to any value of v_i, then μ is uniquely determined by (19) and both x and y vary continuously in Y. If however λ is equal to a value of v_i, then y varies continuously in Y but the solution for x has a discontinuity. Technically speaking, the mapping from values of Y to the corresponding set of solutions for x is upper semi-continuous although not unique at that point (see Exercise 11). If Y is slightly smaller than the critical value, λ is slightly larger than the value of v_i, and not equal to any other value v_j. μ is then unique and approximately equal to the right side of (19): $\mu \doteq \mu_1$. If Y is slightly larger than the critical value, λ is slightly smaller than the value of v_i, but not equal to any other value v_j. μ is then unique and approximately equal to the left side of (19): $\mu \doteq \mu_0$. Thus μ makes a downward leap of magnitude $\mu_1 - \mu_0$ as Y passes up through the critical value. Hence all the x_i with $v_i > \lambda$ jump downward from $\dfrac{\mu_1}{\lambda \alpha_i}$ to $\dfrac{\mu_0}{\lambda \alpha_i}$ and those with $v_i = \lambda$ jump upward from zero to $\dfrac{\mu_0}{\lambda \alpha_i}$.

An application. Suppose the problem (9) resulted from the following model. There are n weapons types from which we have to choose, given a total budget X. After we have chosen a "mixture" $x = (x_1, \ldots, x_n)$, an enemy counters the ith system by the application of y_i units of "counterforce" effort. His budget is taken to be Y. Suppose this leaves a residual quantity of $x_i' = x_i e^{-\alpha_i y_i}$ units of the ith system. The positive numbers α_i are called the "vulnerabilities." Suppose the value of one residual unit of the ith system is v_i, and suppose the values of the various systems add. Then the total residual value is given by:

$$F(x, y) = \Sigma v_i x_i e^{-\alpha_i y_i}.$$

The budget restrictions are $\Sigma x_i = X$, $x_i \geqq 0$ and $\Sigma y_i = Y$, $y_i \geqq 0$.

We thus have a problem of type (9). It has a pure strategy solution in the sense of game theory, so that the Max-Min theory need not yet specially be invoked. The solution obtained above may be interpreted as follows:

(i) The criterion for buying or not buying a weapons system is its unit effectiveness, or value, v_i. There is a critical value λ such that all systems of higher values should be bought, and all systems of lower values rejected. This value λ depends only on the *enemy's* budget.

(ii) Though the criterion for buying or not buying a weapons system is its effectiveness, *once it has met this entrance requirement as specified in* (i), *it is bought in inverse proportion to its vulnerability.*

(iii) In the special case that there are one or more weapons systems of effectiveness equal to the critical effectiveness, there are a number of possible solutions. If the enemy's budget may be somewhat lower than the one being considered, the best solution will be one in which these particular weapons systems are rejected. If it is possibly somewhat higher, the best solution will be one in which these weapons systems are accepted along with those above the critical level, and all in amounts inversely proportional to their vulnerabilities. If there are no such side considerations, these weapons may be rejected or accepted (up to the inverse proportionality limit) as desired.

(iv) The enemy should attack all systems which should be un-ambiguously bought, i.e., have effectiveness *above* the critical level. The amount of attack should be such that the residual value of one unit of initial purchase is be equal to that critical level.

Thus one can go quite a way with problem (9). The principles stated for the above are far from obvious without the application of mathematics. Of course we do not mean the principles to apply to any but this problem. The example worked in Exercises 4—10 violates the three last principles stated above.

Exercises to Chapter II

1. Interpret in verbal terms the solution of problem (9) when it is regarded as the finite search problem at the end of Section 3.

2. Given $a_1 = 1$, $a_2 = 2$, $a_3 = 3$, $b_1 = b_2 = b_3 = 1$, and $X = 1$, solve equation (8) numerically and thus find the explicit values of x_1, x_2, and x_3.

3. Prove that $X(y)$ and $Y(x)$ are convex if $F(x, y)$ is concave-convex (see footnote p. 13).

Exercises 4—10 consider the problem defined by

$$F(x, y) = \Sigma v_i (1 - e^{-\beta_i x_i}) e^{-\alpha_i y_i}$$

subject to $\Sigma x_i = X$, $x_i \geq 0$, $\Sigma y_i = Y$, $y_i \geq 0$. Since $F(x, y)$ is concave in x and convex in y, there is a pure-strategy game solution.

4. Find the necessary conditions for the solution. Use λ for the x-player and μ [after the change of sign as in deriving (14)] for the y-player. Thus the necessary conditions will read $=\lambda$ or $\leq \lambda$ and $=\mu$ or $\leq \mu$.

    5. Prove that $\mu > 0$, and thence that $y_i > 0$ implies $x_i > 0$.

    6. Prove that $x_i > 0$ if and only if $v_i \beta_i > \lambda$.

    7. Show that, if $x_i > 0$, $y_i > 0$ if and only if $v_i \alpha_i (1 - \dfrac{\lambda}{v_i \beta_i}) > \mu$.

    8. Using the result of Exercise 6, write down the formulas for the x_i and y_i. *Crib*: when $y_i > 0$ its formula is

$$y_i = \frac{1}{\alpha_i} \log \frac{\dfrac{v_i \beta_i}{\lambda}}{1 + \dfrac{\mu \beta_i}{\lambda \alpha_i}}.$$

    9. A function $f(x) = f(x_1, \ldots, x_n)$ in a convex set is said to be *convex* if, for any t with $0 \leq t \leq 1$, and for any two points x^0 and x^1,

$$f((1-t)x^0 + tx^1) \leq (1-t)f(x^0) + tf(x^1).$$

If the above inequality holds strictly for $0 < t < 1$ and $x^0 \neq x^1$, the function is called *strictly convex*. Similarly one defines concavity. (See Exercise 6 of Chapter III.) Prove using convexity and Exercise 5 that the solution for y is unique, and using concavity that the solution for x is unique.

    10. (i) Suppose $F(x, y)$ represented a model for attack with residual forces as does the game example in the text. State the analogue of principle (i) of the text.

    (ii) Suppose all the β_i's equal to unity. The condition of Exercise 7 divides the weapon types with $x_i > 0$ into two classes.

    Can a statement analogous to principle (ii) of the text be made for either of these classes?

    (iii) There is no ambiguity in this problem, as Exercise 9 shows. Uniqueness then implies stability (i.e., continuity) with respect to both X and Y. Why?

    (iv) Principle (iv) does not hold for this problem. Recall the reason, which you should have already worked out above.

    11. Let φ be a point-to-set mapping in Euclidean space, so that if u is a point $\varphi(u)$ is a set. Suppose that whenever $u_n \to u_0$, $v_n \to v_0$, and $v_n \in \varphi(u_n)$ for $n = 1, 2, \ldots$, then $v_0 \in \varphi(u_0)$. Then the mapping φ is called *upper semi-continuous*. This concept generalizes that of continuity for a point-to-point mapping. Define a mapping ψ from the real line to n-space as follows: With X fixed, let $\psi(Y)$ be the set of x-solutions to problem (9) of the text, as given by Lemma 9. Prove the statement made in the discussion of stability, that ψ is an upper semi-continuous mapping.

Chapter III

The directional derivative

In this chapter we shall be concerned with a function $F(x, y)$ of two variables x and y. x is supposed to be a point of Euclidean space R^n, $x = (x_1, ..., x_n)$, and y a point of some compact topological space $\mathcal{Y}$. While all the considerations of this chapter are carried out for a general $\mathcal{Y}$, the applications made in this book have $\mathcal{Y}$ a closed bounded set in Euclidean space, and the reader may therefore if he wishes read "closed bounded Euclidean" when he sees "compact topological."

We shall suppose that $F(x, y)$ is continuous in its variables taken together, and also that the partial derivatives $F_{x_i}(x, y)$ are continuous in those variables taken together.

The principal questions at hand concern the function

$$\varphi(x) = \operatorname*{Min}_{y} F(x, y),$$

the minimum being taken over all $y \in \mathcal{Y}$. The example

$$F(x, y) = y \sin x$$

noted in Chapter I, where $-1 \leqq y \leqq +1$, shows that $\varphi(x)$ need not be differentiable in the usual sense. In Chapter I we saw that $\varphi(x) = -|\sin x|$ is not differentiable when $x = 0$. However it does have a right and a left derivative at $x = 0$; and this fact suggests the key to the situation.

At a maximum the derivative of a function, if it exists, need not be zero; it need only be non-positive in every admissible direction. With this in mind we can make a further remark: the function does not need to be differentiable; all that is needed is a *directional* derivative in every admissible direction; then, at a maximum, the necessary condition will be that the directional derivative in any admissible direction is non-positive.

We have left unclear here the meaning of "admissible direction"; this means any direction not pointing out of the space (possibly constrained) of x's admissible for the problem at hand. We shall give a precise definition in the discussion of the Lagrange multiplier principle (Theorem VII) for the case (one constraint) there treated. In the Appendix, with more than one constraint, "admissible" has a somewhat generalized meaning.

2*

For a given point x^0 to yield a maximum it is not even necessary that a directional derivative exist in every admissible direction. It is evident that if x^0 yields a maximum to $\varphi(x)$, then $\varphi(x) \leq \varphi(x^0)$ for all x, so that all the difference quotients

$$\frac{\varphi(x) - \varphi(x^0)}{|x - x^0|}$$

as $x \to x^0$ along any direction must be non-positive. It is therefore natural to inquire how these quotients might relate to the original function $F(x, y)$.

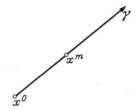

Fig. 2. Illustrating the direction γ from the point x^0

We shall investigate this question. Suppose $x^m \to x^0$ from the direction γ, with $x^m \neq x^0$. By $\gamma = (\gamma_1, \ldots, \gamma_n)$ is meant a vector of length unity pointing away from x^0 and with the direction cosines $\gamma_1, \ldots, \gamma_n$. Thus $\gamma_1^2 + \cdots + \gamma_n^2 = 1$. Alternatively one may think of γ as a point on the unit sphere; the vector from the origin to γ is the direction in question.

If $x^m \to x^0$ from the direction γ, then for all i

$$\frac{x_i^m - x_i^0}{d^m} = \gamma_i \tag{1}$$

where $d^m = |x^m - x^0| = \sqrt{\sum_{i=1}^{n} (x_i^m - x_i^0)^2}$ is the n-dimensional distance from x^m to x^0.

Let $Y(x)$ be the set of those $y \in \mathcal{Y}$ which yield the minimum to $F(x, y)$ for x fixed. It is clear that there always is at least one point $y \in Y(x)$, since $F(x, y)$ is continuous and $\mathcal{Y}$ is compact; see Exercise 1. It is sometimes convenient to use the expression "answering set" for $Y(x)$; the elements y of $Y(x)$ "answer" x.

We now turn to the difference quotient

$$\frac{\varphi(x^m) - \varphi(x^0)}{|x^m - x^0|} = \frac{\varphi(x^m) - \varphi(x^0)}{d^m}. \tag{2}$$

Let $y^m \in Y(x^m)$, $m \geq 1$, and $y^0 \in Y(x^0)$. We are supposing that $x^m \to x^0$. We may rewrite (2) in the form

$$\frac{\varphi(x^m) - \varphi(x^0)}{d^m} = \frac{F(x^m, y^m) - F(x^0, y^0)}{d^m}$$

$$= \frac{F(x^m, y^m) - F(x^m, y^0)}{d^m} + \frac{F(x^m, y^0) - F(x^0, y^0)}{d^m}. \tag{3}$$

Since y^m answers x^m, the first term of (3) is negative or zero. Hence we have from (3)

$$\frac{\varphi(x^m) - \varphi(x^0)}{d^m} \leqq \frac{F(x^m, y^0) - F(x^0, y^0)}{d^m}. \tag{4}$$

We may now apply the law of the mean for a function of n variables to the right side of (4). We get

$$\frac{F(x^m, y^0) - F(x^0, y^0)}{d^m} = \sum_{i=1}^{n} \frac{x_i^m - x_i^0}{d^m} F_{x_i}(x^{m(\theta)}, y^0), \tag{5}$$

where $x^{m(\theta)} = (1 - \theta)x^0 + \theta x^m$ for some θ (depending on m) with $0 < \theta < 1$, i.e., $x^{m(\theta)}$ is some point on the segment joining x^0 to x^m. Using (1) in (5) and combining with (4), we get

$$\frac{\varphi(x^m) - \varphi(x^0)}{d^m} \leqq \sum_{i=1}^{n} \gamma_i F_{x_i}(x^{m(\theta)}, y^0). \tag{6}$$

Now $x^{m(\theta)}$ lies on the segment joining x^0 to x^m, so that $x^{m(\theta)} \to x^0$ as $m \to \infty$. We have assumed that $F_{x_i}(x, y)$ is continuous; hence

$$F_{x_i}(x^{m(\theta)}, y^0) \to F_{x_i}(x^0, y^0)$$

as $m \to \infty$. Thus the right side of (6) tends to a limit equal to

$$\sum_{i=1}^{n} \gamma_i F_{x_i}(x^0, y^0).$$

The ratios on the left side therefore have an upper limit bounded by that quantity as $m \to \infty$:

$$\limsup_{m \to \infty} \left[\frac{\varphi(x^m) - \varphi(x^0)}{d^m} \right] \leqq \sum_{i=1}^{n} \gamma_i F_{x_i}(x^0, y^0). \tag{7}$$

The estimate (7) evidently holds for any sequence $x^m \to x^0$ from the direction γ. Also, y^0 is an arbitrary element of $Y(x^0)$. We may therefore rewrite (7) in the form

$$\limsup_{x^m \to x^0} \left[\frac{\varphi(x^m) - \varphi(x^0)}{d^m} \right] \leqq \operatorname*{Min}_{y \in Y(x^0)} \sum_{i=1}^{n} \gamma_i F_{x_i}(x^0, y), \tag{8}$$

understanding that (8) holds for *any* sequence $x^m \to x^0$ from the direction γ with $x^m \neq x^0$.

Now we consider the infimum of the difference quotient (2). Suppose that $m^* \to \infty$ and that

$$\lim_{m^* \to \infty} \left[\frac{\varphi(x^{m^*}) - \varphi(x^0)}{d^{m^*}} \right] = \liminf_{x^m \to x^0} \left[\frac{\varphi(x^m) - \varphi(x^0)}{d^m} \right]. \tag{9}$$

Let $y^{m^*} \in Y(x^{m^*})$. For a subsequence $m' \to \infty$ the elements $y^{m'}$ converge to a limit, which we denote by y^γ. $y^\gamma \in Y(x^0)$ because $F(x, y)$ is continuous in the variables x and y taken together (Exercise 2). Hence

$$\frac{\varphi(x^{m'}) - \varphi(x^0)}{d^{m'}} = \frac{F(x^{m'}, y^{m'}) - F(x^0, y^\gamma)}{d^{m'}}$$

$$= \frac{F(x^{m'}, y^{m'}) - F(x^0, y^{m'})}{d^{m'}} + \frac{F(x^0, y^{m'}) - F(x^0, y^\gamma)}{d^{m'}}$$

$$\geqq \frac{F(x^{m'}, y^{m'}) - F(x^0, y^{m'})}{d^{m'}}$$

$$= \sum_{i=1}^{n} \gamma_i F_{x_i}(x^{m'(\omega)}, y^{m'}),$$

where $x^{m'(\omega)} = (1 - \omega)x^0 + \omega x^{m'}$ with $0 < \omega < 1$ (ω depending on m') is some point on the segment joining x^0 to $x^{m'}$. Again $x^{m'(\omega)} \to x^0$, so that on passing to the limit we find that

$$\lim_{m' \to \infty} \left[\frac{\varphi(x^{m'}) - \varphi(x^0)}{d^{m'}} \right] \geqq \sum_{i=1}^{n} \gamma_i F_{x_i}(x^0, y^\gamma). \tag{10}$$

Recalling that $\{m'\}$ is a subsequence of $\{m^*\}$ and referring to the definition (9), we thus obtain

$$\liminf_{x^m \to x^0} \left[\frac{\varphi(x^m) - \varphi(x^0)}{d^m} \right] \geqq \sum_{i=1}^{n} \gamma_i F_{x_i}(x^0, y^\gamma) \tag{11}$$

for the element $y^\gamma \in Y(x^0)$, which appeared as a limit of answers y^m to elements $x^m \to x^0$ from the direction γ.

Putting (8) and (11) together, we arrive at the fundamental theorem of the theory of Max-Min, which follows.

Theorem I.[5] *Let* $\gamma_1, \ldots, \gamma_n$, *where* $\gamma_1^2 + \cdots + \gamma_n^2 = 1$, *denote any direction. Then the directional derivative* $D_\gamma \varphi(x)$ *of* φ *in the direction* γ *at the point* x *exists, and is given by the formula*

$$D_\gamma \varphi(x) = \underset{y \in Y(x)}{\text{Min}} \sum_{i=1}^{n} \gamma_i F_{x_i}(x, y), \tag{12}$$

---

[5] The author wishes to call attention to a somewhat similar theorem, on derivative of the value of a game, obtained by OLIVER GROSS in 1954 and referred to in a paper by HARLAN MILLS in [7]. These results, which apparently received at the time no further development, were not known to this author. See also the Addendum to [8].

where $Y(x)$ denotes the set of elements $y \in \mathcal{Y}$ minimizing against the vector x. Further, let $\{x^m\}$ be any sequence approaching x along the vector of direction γ issuing from x, and let $y^m \in Y(x^m)$. Then there is a subsequence $m' \to \infty$ such that the sequence $\{y^{m'}\}$ converges to an element $y^{\gamma} \in Y(x)$ which yields the minimum in (12).

Remark. *Arcs with direction γ.* Consider any arc issuing from x. Let x^m be any sequence of points tending to x along the arc, and put

$$\gamma_i^m = \frac{x_i^m - x_i}{d^m},$$

d^m being the distance from x^m to x. If

$$\lim_{m \to \infty} \gamma_i^m = \gamma_i$$

for any such sequence, then the arc is said to issue from x with direction γ. It is evident that Theorem I holds with this generalized interpretation of direction; thus if the x^m tend to x along any arc with direction γ issuing from x, the difference quotients approach $D_\gamma \varphi(x)$ (Exercise 3). This remark is especially important in dealing with non-linear constraints. See for instance the Appendix.

Theorem I is a generalization of a trivial formula of classical mathematics. Let $f(x, y)$ be a twice continuously differentiable function on the unit square, with $f_{yy}(x, y) > 0$ everywhere. Suppose that for each x the minimum $\varphi(x)$ of $f(x, y)$ is taken on at an interior value of y. For that value $f_y(x, y) = 0$, and since $f_{yy}(x, y) > 0$ the y is unique. By the implicit function theorem ([3], p. 138) one may in fact solve for y yielding the minimum as a continuously differentiable function of $x: y = y(x)$. Hence

$$\varphi(x) = f(x, y(x)),$$

so that $\varphi(x)$ is a continuously differentiable function of x. The derivative is

$$\varphi'(x) = f_x(x, y(x)) + f_y(x, y(x)) \cdot y'(x)$$
$$= f_x(x, y(x)) \tag{13}$$

since $f_y(x, y(x)) \equiv 0$. That (13) is a special case of the general formula (12) is evident.

The reader should now review the example $F(x, y) = y \sin x$ worked out in Chapter I. Notice that the only value of x where the ordinary derivative of $\varphi(x)$ does not exist is $x = 0$, and that for that value of x the solution for y is not unique; in fact it is the whole interval $[-1, +1]$. The connection between these facts is clear from formula (12). If $Y(x)$ consists of a single point y we have

$$D_\gamma \varphi(x) = \Sigma \gamma_i F_{x_i}(x, y) \tag{14}$$

for every γ. In particular, if we denote by $D_{i+}\varphi(x)$ the derivative in the positive x_i direction and by $D_{i-}\varphi(x)$ the derivative in the negative x_i direction, we find from (14)

$$D_{i+}\varphi(x) = F_{x_i}(x, y) \tag{15}$$

and

$$D_{i-}\varphi(x) = -F_{x_i}(x, y). \tag{16}$$

(15) is the right derivative of $\varphi(x)$ with respect to x_i, and (16) the negative of the left derivative of $\varphi(x)$ with respect to x_i. Thus the left and right derivative of $\varphi(x)$ with respect to x_i exist and are both equal to $F_{x_i}(x, y)$. It follows that the usual partial derivative of $\varphi(x)$ with respect to x_i exists and is equal to $F_{x_i}(x, y)$. Using the notation $\varphi_i(x)$ to denote this partial derivative, we may rewrite (14) in the form

$$D_\gamma\varphi(x) = \Sigma\gamma_i\varphi_i(x).$$

This is the same as the elementary rule of calculus for calculating the directional derivative.

Now suppose in addition that the solution y is unique throughout a neighborhood of x. Then y is a continuous function of x in the same neighborhood [see Exercise 10 (iii) of Chapter II]. Denoting this solution by $y(x)$, we find from (15) and (16) that

$$\varphi_i(x) = F_{x_i}(x, y(x))$$

is a continuous function of x. Assembling the facts enumerated in these two paragraphs, we arrive at the following theorem.

Theorem II. *Suppose for some point x that the minimizing set $Y(x)$ consists of a single point $y(x)$. Then the usual partial derivatives $\varphi_i(x) = \partial\varphi/\partial x_i$ of $\varphi(x)$ exist at the point x, and the directional derivative is given by the "chain rule" of the elementary calculus:*

$$D_\gamma\varphi(x) = \Sigma\gamma_i\varphi_i(x). \tag{17}$$

If furthermore the solution $y(x)$ is unique throughout a neighborhood, then the partial derivatives of $\varphi(x)$ are continuous in that neighborhood, i.e. $\varphi(x)$ is C^1 there.

The "seesaw" example has shown that the situation described by Theorem II is by no means general. We now present another example of the breakdown of the "chain rule." Let the spaces for x and y be as follows: $x = (x_1, x_2)$ with $x_1, x_2 \geqq 0$, and $y = (y_1, y_2)$ with $y_1, y_2 \geqq 0$ and $y_1 + y_2 = \dfrac{\pi}{2}$. Put

$$F(x, y) = F(x_1, x_2, y_1, y_2) = x_1 \sin y_1 + x_2 \sin y_2. \tag{18}$$

Elementary calculus arguments (see Exercises 5 and 6) show that for fixed x_1, x_2 the minimum to $F(x, y)$ must be either at $y = (0, \pi/2)$ or

$y = (\pi/2, 0)$. Referring to (18) we find immediately that

$$\varphi(x) = \varphi(x_1, x_2) = \operatorname*{Min}_{i=1,2} x_i. \qquad (19)$$

Let us plot $\varphi(x)$ as a function of x_1 for $x_2 = 1/2$.

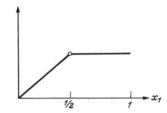

Fig. 3. Plot of $\varphi(x_1, x_2)$ as a function of x_1 for $x_2 = 1/2$

Write $\gamma^1 = (1, 0)$ and $\gamma^2 = (0, 1)$. The right derivative in x_1 of $\varphi(x^1, 1/2)$ is seen from the figure to be zero; thus

$$D_{\gamma^1}\varphi(1/2, 1/2) = 0. \qquad (20)$$

Similarly

$$D_{\gamma^2}\varphi(1/2, 1/2) = 0. \qquad (21)$$

Now write $\gamma^* = (1/\sqrt{2}, 1/\sqrt{2})$. It is obvious from (19) (Exercise 7) that

$$D_{\gamma^*}\varphi(1/2, 1/2) = \frac{1}{\sqrt{2}} > 0. \qquad (22)$$

Now

$$\gamma^* = \frac{1}{\sqrt{2}}\gamma^1 + \frac{1}{\sqrt{2}}\gamma^2. \qquad (23)$$

If the elementary rule for computing the directional derivative were to hold we would have from (23) that

$$D_{\gamma^*}\varphi(1/2, 1/2) = \frac{1}{\sqrt{2}}D_{\gamma^1}\varphi(1/2, 1/2) + \frac{1}{\sqrt{2}}D_{\gamma^2}\varphi(1/2, 1/2). \qquad (24)$$

But the left side of (24) is positive and the right side, from (20) and (21), zero.

Thus the elementary "chain" rule does not hold in general. This fact is perhaps the most important feature of the theory: *it means that calculations with directional derivatives cannot be reduced to differentiations along the coordinate axes.* In the elementary calculus with smooth functions in R^n, one is concerned with the n partial derivatives at each point. In the Max-Min theory we must be concerned with infinitely many derivatives at each point.

There is however a partial replacement for the "chain rule." This is the following superadditivity rule.

Theorem III. *Let α, β, and γ be directions satisfying $\gamma = a\alpha + b\beta$ for some a, $b > 0$. Then*

$$D_\gamma \varphi(x) \geqq a D_\alpha \varphi(x) + b D_\beta \varphi(x). \tag{25}$$

Proof. The proof is trivial:

$$
\begin{aligned}
D_\gamma \varphi(x) &= \operatorname*{Min}_{y \in Y(x)} \Sigma \gamma_i F_{x_i}(x, y) \\
&= \operatorname*{Min}_{y \in Y(x)} \left[a \Sigma \alpha_i F_{x_i}(x, y) + b \Sigma \beta_i F_{x_i}(x, y) \right] \\
&\geqq a \operatorname*{Min}_{y \in Y(x)} \Sigma \alpha_i F_{x_i}(x, y) + b \operatorname*{Min}_{y \in Y(x)} \Sigma \beta_i F_{x_i}(x, y) \\
&= a D_\alpha \varphi(x) + b D_\beta \varphi(x).
\end{aligned}
$$

That the directional derivative in a given direction γ is not continuous as a function of x is shown by the seesaw example. However the following theorems hold.

Theorem IV. $D_\gamma \varphi(x)$ *is lower semicontinuous in the variables γ and x taken together.*

Proof. Suppose that $x^m \to x^0$ and $\gamma^m \to \gamma^0$. We have to show that

$$\liminf_{m \to \infty} D_{\gamma^m} \varphi(x^m) \geqq D_{\gamma^0} \varphi(x^0). \tag{26}$$

Take a subsequence $m' \to \infty$ so that

$$\lim_{m' \to \infty} D_{\gamma^{m'}} \varphi(x^{m'}) = \liminf_{m \to \infty} D_{\gamma^m} \varphi(x^m). \tag{27}$$

We have, for some $y^{m'} \in Y(x^{m'})$,

$$D_{\gamma^{m'}} \varphi(x^{m'}) = \Sigma \gamma_i^{m'} F_{x_i}(x^{m'}, y^{m'}). \tag{28}$$

We choose a subsequence $m^* \to \infty$ of the sequence $m' \to \infty$ so that y^{m^*} converges to a limit y^*. Evidently $y^* \in Y(x^0)$. Because $F_{x_i}(x, y)$ is continuous in its variables taken together, the limit on the right side of (28) is

$$\lim_{m^* \to \infty} \Sigma \gamma_i^{m^*} F_{x_i}(x^{m^*}, y^{m^*}) = \Sigma \gamma_i^0 F_{x_i}(x^0, y^*).$$

Combining this with (27) we obtain

$$\liminf_{m \to \infty} D_{\gamma^m} \varphi(x^m) = \Sigma \gamma_i^0 F_{x_i}(x^0, y^*). \tag{29}$$

But since $y^* \in Y(x^0)$,

$$\Sigma \gamma_i^0 F_{x_i}(x^0, y^*) \geqq \operatorname*{Min}_{y \in Y(x^0)} \Sigma \gamma_i^0 F_{x_i}(x^0, y) = D_{\gamma^0} \varphi(x^0). \tag{30}$$

(29) and (30) imply (26), and the theorem is proved.

Theorem V. $D_\gamma \varphi(x)$ *is continuous as a function of γ alone.*

Proof. From Theorem IV, we need only prove upper semi-continuity.[6] Suppose that $y^0 \in Y(x^0)$ yields the minimum in the formula for $D_{\gamma^0}\varphi(x^0)$. Let $\gamma^m \to \gamma^0$. We have

$$D_{\gamma^m}\varphi(x^0) = \operatorname*{Min}_{y \in Y(x^0)} \Sigma \gamma_i^m F_{x_i}(x^0, y) \leqq \Sigma \gamma_i^m F_{x_i}(x^0, y^0).$$

Hence

$$\limsup_{m \to \infty} D_{\gamma^m}\varphi(x^0) \leqq \limsup_{m \to \infty} \Sigma \gamma_i^m F_{x_i}(x^0, y^0)$$

$$= \Sigma \gamma_i^0 F_{x_i}(x^0, y^0) = D_{\gamma^0}\varphi(x^0)$$

and the theorem is proved.

We now give the analogue to the law of the mean. We preface it with a trivial lemma about functions on $[0, 1]$.

Lemma. *Suppose $f(x)$ is continuous on $[0, 1]$ and has $f(0) = 0$. Suppose that $f(x)$ has a non-negative right derivative $f_R(x)$ at every point of $[0, 1]$. Then $f(1) \geqq 0$. If $f_R(x) > 0$ at any point, $f(1) > 0$.*

Proof. Consider the first statement of the lemma. Suppose $f(1) = -2a < 0$. Put $g(x) = f(x) + ax$. Then $g_R(x) = f_R(x) + a \geqq a$ for all $x \in [0, 1)$. Let x^* yield the maximum to $g(x)$ in $[0, 1]$. Suppose $x^* < 1$. Then $g_R(x^*) \geqq a > 0$, so that there are points to the right of x^* with $g(x) > g(x^*)$. Hence $x^* = 1$, and in particular $g(1) \geqq 0$. Hence $f(1) \geqq -a$, which contradicts the definition $f(1) = -2a$, and the first statement of the lemma is proved. The second statement is an easy consequence of the first statement (Exercise 8).

Theorem VI.(Law of the mean). *Let γ be the direction of the segment joining x^0 to x^1. Then there exists a point $x \neq x^1$ on that segment at which*

$$D_\gamma \varphi(x) \leqq \frac{\varphi(x^1) - \varphi(x^0)}{|x^1 - x^0|} \tag{31}$$

and a point $x' \neq x^1$ at which

$$D_\gamma \varphi(x') \geqq \frac{\varphi(x^1) - \varphi(x^0)}{|x^1 - x^0|}. \tag{32}$$

Proof. Put $x^\theta = (1 - \theta)x^0 + \theta x^1$, where $0 \leqq \theta \leqq 1$. Put

$$\psi(\theta) = \varphi(x^\theta) - \varphi(x^0) - \theta[\varphi(x^1) - \varphi(x^0)].$$

Then

$$\psi(0) = \psi(1) = 0. \tag{33}$$

Secondly, the right derivative of ψ with respect to θ exists at every point with $0 \leqq \theta < 1$ and is given by (Exercise 10)

$$\psi_R(\theta) = |x^1 - x^0| D_\gamma \varphi(x^\theta) - [\varphi(x^1) - \varphi(x^0)]. \tag{34}$$

Now if $\psi_R(\theta) > 0$ throughout $[0, 1)$, we would have from the lemma that $\psi(1) > \psi(0)$, contradicting (33). It follows that $\psi_R(\theta) \leqq 0$ for some

---

[6] The fact that $D_\gamma \varphi(x)$ is upper semi-continuous as a function of γ alone was pointed out by J. BRAM.

$\theta < 1$, from which we immediately obtain (31). As to (32), it follows from the fact that $\psi_R(\theta') \geqq 0$ for some $\theta' < 1$.

The Lagrange Multiplier

The final topic of this chapter is the Lagrange multiplier theorem. Most proofs of the classical Lagrange multiplier theorem make use of differentiation with respect to the directions of the axes. It therefore might seem to be too much to ask to expect a Lagrange multiplier theorem for the directional derivative of this chapter, which cannot be reduced in general to derivatives along the axes. On the other hand, the Lagrange principle lies very deep and appears in many guises in many parts of analysis. It is worthwhile considering what it would mean in the type of problem that interests us.

First a remark about the classical constrained problem, without edges. What the classical Lagrange multiplier theorem does is the following. Given that a function has a critical point subject to a constraint, the Lagrange theorem states that there is assigned to the function and the constraint another function, and then that the new function has a critical point at the same point, this time in the unconstrained space. *Thus what that theorem does is to provide information concerning the neighboring space, information given through the behavior of the new function.*

Let us try to apply this line of thought to the problem of maximizing $\varphi(x)$ subject to the simplest possible constraint:

$$\Sigma x_i = X, \quad x_i \geqq 0. \tag{35}$$

Suppose x^0 yields the maximum to $\varphi(x)$ subject to these constraints.

We need two definitions related to x^0. The vector $\gamma = (\gamma_1, \ldots, \gamma_n)$ is *possible* if $\gamma_i \geqq 0$ for these values of i with $x_i^0 = 0$. It is *admissible* if it is possible and $\Sigma \gamma_i = 0$. A vector issuing from x^0 is then possible if it does not point strictly out of the positive orthant $x_1 \geqq 0, \ldots, x_n \geqq 0$. It is admissible if it lies as well in the plane $\Sigma x_i = X$. In fact, starting from x^0 and going a distance $\varepsilon > 0$ along an admissible vector γ will lead to the point $x_1^0 + \varepsilon \gamma_1, \ldots, x_n^0 + \varepsilon \gamma_n$. This point has

$$\sum_{i=1}^n (x_i^0 + \varepsilon \gamma_i) = \sum_{i=1}^n x_i^0 + \varepsilon \sum_{i=1}^n \gamma_i = X,$$

and therefore it satisfies the constraint as well.

If then x^0 yields the maximum subject to (35), it is obviously a necessary condition that

$$D_\gamma \varphi(x^0) \leqq 0 \tag{36}$$

for every admissible direction γ.

In following the line of thought begun above, we seek information about the unconstrained space neighboring x^0. We make two more definitions: a possible vector γ is an *up* vector if $\Sigma \gamma_i > 0$ and a *down* vector if $\Sigma \gamma_i < 0$. The method of obtaining the desired information is the following. We seek to combine up and down vectors so as to get admissible vectors. Thus, suppose that $\Sigma \gamma_i = a > 0$ and $\Sigma \gamma_i' = -b < 0$. Suppose for the moment that $\gamma + \gamma' \neq 0$. We make an admissible vector out of γ and γ' by writing

$$\gamma'' = \delta(b\gamma + a\gamma'). \tag{37}$$

Here $\delta > 0$ is chosen so as to make γ'' a unit vector. This is possible since if $b\gamma + a\gamma'$ were the zero vector, then $b\gamma_i + a\gamma_i' = 0$ term by term, i.e. $\gamma_i' = -\dfrac{b}{a}\gamma_i$. Since both are unit vectors we must then have $b = a$, so that $\gamma_i' = -\gamma_i$ for all i, the case excluded above.

Thus using (37) it is possible to combine an up vector with a down vector not directly opposed to it so as to get a vector γ'' which is admissible, since

$$\Sigma \gamma_i'' = \delta b \Sigma \gamma_i + \delta a \Sigma \gamma_i' = \delta ba - \delta ab = 0.$$

The reader is advised to draw a picture.

Now (36) says that $D_{\gamma''}\varphi(x^0) \leq 0$. At the same time Theorem III [formula (25)] states that, because of formula (37),

$$D_{\gamma''}\varphi(x^0) \geq \delta b D_\gamma \varphi(x^0) + \delta a D_{\gamma'}\varphi(x^0). \tag{38}$$

It follows from (38) and (36) that

$$b D_\gamma \varphi(x^0) + a D_{\gamma'}\varphi(x^0) \leq 0,$$

or, recalling the definitions of a and b,

$$\frac{D_\gamma \varphi(x^0)}{\Sigma \gamma_i} \leq \frac{D_{\gamma'}\varphi(x^0)}{\Sigma \gamma_i'}. \tag{39}$$

Thus by forming an admissible vector from the up vector γ and the down vector γ', and applying the fact that x^0 maximizes subject to the constraint, we have arrived at an inequality referring to the up and down vectors only. This gives us the desired information about the unconstrained space neighboring x^0.

Before proceeding, we must deal with the case when $\gamma + \gamma' = 0$. In this case take a sequence γ^m approaching γ and with $\gamma^m \neq \gamma$. Then $\gamma^m + \gamma' \neq 0$, so that (39) may be applied:

$$\frac{D_{\gamma^m}\varphi(x^0)}{\Sigma \gamma_i^m} \leq \frac{D_{\gamma'}\varphi(x^0)}{\Sigma \gamma_i'}. \tag{40}$$

The denominator on the left side of (40) is bounded. And, from Theorem IV, $D_\gamma\varphi(x)$ is continuous as a function of γ alone. Hence we may pass to the limit on the left side of (4) as $\gamma^m \to \gamma$. The result is that (39) holds in the case $\gamma + \gamma' = 0$ as well.

We now are almost finished. Corresponding to the set of all up vectors we have a collection of numbers

$$\frac{D_\gamma\varphi(x^0)}{\Sigma\gamma_i}, \tag{41}$$

which we shall denote by L. Corresponding to the set of all down vectors we get a collection of numbers (41), which we shall denote by R. The inequality (39) states that if $u \in L$ and $v \in R$ then $u \leq v$. This is exactly the hypothesis for the Dedekind cut (see [3], pp. 28—29): given any two sets L and R with the above property, then there is a (at least one) number λ separating them, so that $u \leq \lambda$ for every element u of L and $v \geq \lambda$ for every element v of R. Hence: if γ is an up vector we have

$$\frac{D_\gamma\varphi(x^0)}{\Sigma\gamma_i} \leq \lambda,$$

so that

$$D_\gamma\varphi(x^0) \leq \lambda\Sigma\gamma_i. \tag{42}$$

If γ' is a down vector we have

$$\frac{D_{\gamma'}\varphi(x^0)}{\Sigma\gamma_i'} \geq \lambda,$$

and, multiplying through by the negative denominator on the left, we get

$$D_{\gamma'}\varphi(x^0) \leq \lambda\Sigma\gamma_i'$$

for down vectors as well. Thus (42) holds for all up vectors *and* all down vectors. As to admissible vectors with $\Sigma\gamma_i = 0$, (42) then reads $D_\gamma\varphi(x^0) \leq 0$, which is indeed true because of the original necessary conditions (36).

We have therefore arrived at a Lagrange multiplier theorem by following a very simple and geometrical line of thought. The theorem is the following.

Theorem VII. (Lagrange Multiplier Principle, one side condition). *If x^0 maximizes φ subject to $\Sigma x_i = X$, $x_i \geq 0$, then there exists a λ such that*

$$D_\gamma\varphi(x^0) \leq \lambda\Sigma\gamma_i \tag{43}$$

for any direction γ which does not violate the requirements that $\gamma_i \geq 0$ when $x_i^0 = 0$.

Put in another way, Theorem VII states that if $\varphi(x)$ is maximized subject to $\Sigma x_i = X$, $x_i \geqq 0$, then the function

$$\psi(x) = \varphi(x) - \lambda \Sigma x_i \qquad (44)$$

has a non-positive directional derivative, i.e.

$$D_\gamma \psi(x) \leqq 0$$

in every possible direction. Thus we have come back to the point from which the search for the Lagrange theorem started.

The Lagrange multiplier λ is not unique in general (nor is it always for simple maximum problems; see Exercise 12).

With more than one constraint or with non-linear constraints the situation is much more intricate. Several years ago KUHN and TUCKER treated in [6] the case of several inequality side conditions and a smooth function $f(x_1, \ldots, x_n)$ being maximized. After Theorem VII was proved, JOSEPH BRAM was able to do for the function $\varphi(x_1, \ldots, x_n)$ what KUHN and TUCKER had done for the smooth function $f(x_1, \ldots, x_n)$. His theorem is the following:

Theorem VIII. (BRAM). *If x^0 maximizes $\varphi(x)$ subject to the constraint set*

$$g_1(x_1, \ldots, x_n) \leqq 0,$$
$$\begin{array}{cc} \cdot & \cdot \\ \cdot & \cdot \\ \cdot & \cdot \end{array} \qquad (45)$$
$$g_m(x_1, \ldots, x_n) \leqq 0,$$

the g_j being continuously differentiable and with non-vanishing gradients, and if the Kuhn-Tucker constraint qualification ([6], p. 483) is satisfied, then there exists a set of non-negative numbers $\lambda_1, \ldots, \lambda_m$ such that for any possible direction

$$D_\gamma \varphi(x^0) \leqq \sum_{j=1}^{m} \lambda_j D_\gamma g_j(x^0), \qquad (46)$$

where $D_\gamma g_j(x^0)$ denotes the directional derivative of g_j in the direction γ at x^0, given by the usual formula

$$D_\gamma g_i(x^0) = \sum_{i=1}^{n} \gamma_i \frac{\partial}{\partial x_i} [g_j(x_1^0, \ldots, x_n^0)]. \qquad (47)$$

His proof will be found in the Appendix.

We have thus constructed for the function $\varphi(x) = \underset{y \in Y}{\text{Min}} F(x, y)$ a complete calculus analogous to elementary calculus. It is interesting to note that the Lagrange multiplier theorem is not as useful as it is in the elementary case; a great deal of information is lost in passing from the constrained to the unconstrained space. We shall frequently have to differentiate in odd directions to get this information.

Exercises to Chapter III

1. Prove that a continuous function on a compact space has a minimum. For "compact", non-topological readers may substitute "bounded closed Euclidean."

2. Let $F(x, y)$ be continuous in both variables x, y. Define $Y(x)$ as in the text: $y \in Y(x)$ if y yields the minimum against x. Suppose that $x^m \to x^0$, $y^m \in Y(x^m)$, and $y^m \to y^0$. Prove that $y^0 \in Y(x^0)$, i.e. y^0 minimizes against x^0.

3. Prove that Theorem I holds with the generalized definition of direction given in the Remark following Theorem I.

4. In Exercise 11 to Chapter II we defined upper semi-continuity for a point-to-set mapping. Prove that the mapping $x \to Y(x)$ defined as in the text preceding formula (2) is an upper semi-continuous point-to-set mapping (see also Exercise 2).

5. Write out the elementary calculus argument showing that the minimum to (18) for fixed (x_1, x_2) with $x_1 > 0$, $x_2 > 0$ under the constraints $y_1, y_2 \geq 0$ and $y_1 + y_2 = \pi/2$ is at a corner.

6. A function $f(x) = f(x_1, \ldots, x_n)$ in a convex set is said to be *concave* if, for any t with $0 \leq t \leq 1$, and for any two points x^0 and x^1,

$$f((1 - t) x^0 + t x^1) \geq (1 - t) f(x^0) + t f(x^1).$$

If the above inequality holds strictly for $0 < t < 1$ and $x^0 \neq x^1$, the function is called *strictly concave*. A point x is a corner point in the convex set provided it cannot be written in the form

$$x = \frac{x^0 + x^1}{2}$$

with $x^0 \neq x^1$.

Show that any minimum for a strictly concave function on a convex set must be at a corner point. Apply this general principle to formula (18).

7. Calculate $D_{y*}\varphi(1/2, 1/2)$ for the function (18) directly, using (19), and then from formula (12).

8. Prove the second statement of the lemma preceding Theorem VI.

9. Prove the following lemma analogous to the lemma preceding Theorem VI. If $f(x)$ is continuous on $[0, 1]$ and has a left derivative $f_L(x)$ satisfying $f_L(x) \geq 0$ at every point $x \in (0, 1]$, and if $f(0) = 0$, then $f(1) \geq 0$.

10. Prove formula (34) of the text.

11. ROLLE's theorem (used in the proof of the law of the mean in elementary calculus) does not hold for the function φ. Give an example.

12. Give an example of a simple maximum problem for which the Lagrange multiplier is not unique.

Chapter IV

Some Max-Min examples

This chapter illustrates the application of the directional derivative to problems ranging from trivial to quite difficult.

In discussions of incomplete solutions, it should always be understood that the important problem is the "outside" one.

1. A trivial application

A very simple example of a Max-Min problem is a problem already mentioned in Chapter III. Put

$$F(x, y) = F(x_1, x_2, y_1, y_2) = x_1 \sin y_1 + x_2 \sin y_2 . \tag{1}$$

Suppose $x_1, x_2, y_1, y_2 \geq 0$ and that the restrictions on x and y are $x_1 + x_2 = 1$, $y_1 + y_2 = \pi/2$ respectively. If we write as usual

$$\varphi(x) = \varphi(x_1, x_2) = \operatorname*{Min}_y F(x, y),$$

the minimum being taken over the space $y_1, y_2 \geq 0$, $y_1 + y_2 = \pi/2$, we may apply formula (12) of Chapter III to calculate $D_\gamma \varphi(x)$ for any direction $\gamma = (\gamma_1, \gamma_2)$ in the x-space:

$$D_\gamma \varphi(x) = \operatorname*{Min}_{y \in Y(x)} \Sigma \gamma_i F_{x_i}(x, y), \tag{2}$$

where $Y(x)$ is the set of y yielding $\operatorname*{Min}_y F(x, y)$. Since $F(x, y)$ is for any x strictly concave in y, the solution to the problem $\operatorname*{Min}_y F(x, y)$ is on a corner. Thus, if $x_1 > x_2$, $Y(x)$ consists of the single point $(0, \pi/2)$; if $x_1 < x_2$, $Y(x)$ consists of the single point $(\pi/2, 0)$; and, if $x_1 = x_2$, $Y(x)$ consists of both of these points.

What we seek is the x yielding $\operatorname*{Max}_x \operatorname*{Min}_y F(x, y)$, i.e. $\operatorname*{Max}_x \varphi(x)$.

At such a point x it is evidently necessary that

$$D_\gamma \varphi(x) \leq 0 \tag{3}$$

in any direction γ that is admissible in the sense of Chapter III, i.e., which has $\gamma_i \geq 0$ if $x_i = 0$, and $\gamma_1 + \gamma_2 = 0$. Suppose that $x_1 > x_2$. Then $\gamma = (-1/\sqrt{2}, 1/\sqrt{2})$ is an admissible direction; $Y(x)$ is the point $(0, \pi/2)$, and formula (2) yields

$$\sqrt{2} D_\gamma \varphi(x) = -\sin y_1 + \sin y_2 = 1 . \tag{4}$$

Hence evidently (3) cannot hold if $x_1 > x_2$. The same situation occurs if $x_2 > x_1$. Thus $x_1 = x_2 = 1/2$ is the solution.

2. A finite allocation game

In Chapter II we worked out the problem

$$F(x, y) = \Sigma v_i x_i e^{-\alpha_i y_i}, \tag{5}$$

where $x_i, y_i \geqq 0$, $\Sigma x_i = X$ and $\Sigma y_i = Y$. We shall work this out using the directional derivative. Fix on some x with $\Sigma x_i > 0$. Since $F(x, y)$ is convex in y, the solution for $Y(x)$ consists of a convex set. Using the Gibbs lemma, we find that an element of this convex set must satisfy, for some μ,

$$v_i x_i \alpha_i e^{-\alpha_i y_i} = \mu \quad \text{if} \quad y_i > 0 ;$$

$$\leqq \mu \quad \text{if} \quad y_i = 0. \tag{6}$$

Since some $x_i > 0$, μ must be positive. Hence y_i is positive only when x_i is positive. It follows that the solution for $Y(x)$ is unique (Exercise 2). But now we may apply Theorem II of Chapter III: the function $\varphi(x) = \underset{y}{\text{Min}}\, F(x, y)$ is continuously differentiable as a function of x in the ordinary sense and its partial derivatives are given by

$$\varphi_i(x) = F_{x_i}(x, y) = v_i e^{-\alpha_i y_i}.$$

But now the Gibbs lemma [or rather its trivial generalization to the case of a function $f(x)$ not representable as a sum of functions $f_i(x_i)$] applies and we may write for some λ

$$v_i e^{-\alpha_i y_i} = \lambda \quad \text{if} \quad x_i > 0 ;$$

$$\leqq \lambda \quad \text{if} \quad x_i = 0. \tag{7}$$

We have therefore arrived at the same set of conditions as (14) and (15) in Chapter II, which were obtained there from the saddle-point conditions (12) and (13). The rest of the solution therefore goes through as before.

3. An allocation model with Min Max > Max Min

Consider the problem

$$F(x, y) = \Sigma v_i [1 - e^{-\beta_i x_i e^{-\alpha_i y_i}}] \tag{8}$$

with $\Sigma x_i = X$, $\Sigma y_i = Y$, $x_i, y_i \geqq 0$. All the v_i, α_i, β_i are positive. Seen as a function of x with y fixed, $F(x, y)$ is strictly concave. However the function, with x_i fixed,

$$g_i(y_i) = 1 - e^{-\beta_i x_i e^{-\alpha_i y_i}}, \tag{9}$$

is convex in y_i if $\beta_i x_i\, e^{-\alpha_i y_i} < 1$ and concave in y_i if $\beta_i x_i e^{-\alpha_i y_i} > 1$ (Exercise 3). Thus in particular if $\beta_i X < 1$ for $i = 1, \ldots, n$, (8) is a game with a pure-strategy solution and may be worked by elementary means (Exercise 4). But in general this is not so, as we show by a numerical example below, and (8) must be studied by Max-Min techniques.

We shall study both $\underset{y}{\text{Min}}\, \underset{x}{\text{Max}}\, F(x, y)$ and $\underset{x}{\text{Max}}\, \underset{y}{\text{Min}}\, F(x, y)$ for (8). We shall find that the corresponding criteria, (14) and (21), for x_i to be positive, are in appearance the same for both problems. And if $\beta_i X < 1$ they are indeed the same. But they are not always the same. We present an example, with $\beta_i X > 1$, in which the λ's which appear in (14) and (21) are distinct, and $\underset{x}{\text{Max}}\, \underset{y}{\text{Min}}\, F(x, y) < \underset{y}{\text{Min}}\, \underset{x}{\text{Max}}\, F(x, y)$.

We begin with the Min-Max problem. Define

$$\psi(y) = \underset{x}{\text{Max}}\, F(x, y), \tag{10}$$

where x is subjected to the condition $\Sigma x_i = X$, $x_i \geq 0$. Since, for fixed y, $F(x, y)$ is strictly concave in x, the solution for x in (10) is unique. By the Gibbs lemma it satisfies for some λ (which depends on y) the condition

$$v_i \beta_i e^{-\alpha_i y_i} e^{-\beta_i x_i e^{-\alpha_i y_i}} = \lambda \quad \text{if} \quad x_i > 0\,; $$
$$\leq \lambda \quad \text{if} \quad x_i = 0\,. \tag{11}$$

Applying Theorems I and II of Chapter III, we see that $\psi(y)$ is continuously differentiable as a function of y, and that the partial derivatives are

$$\psi_i(y) = -\, v_i \alpha_i \beta_i x_i e^{-\alpha_i y_i} e^{-\beta_i x_i e^{-\alpha_i y_i}}. \tag{12}$$

Applying the (generalized as in the first example) Gibbs lemma we obtain from (12) the condition

$$v_i \alpha_i \beta_i x_i e^{-\alpha_i y_i} e^{-\beta_i x_i e^{-\alpha_i y_i}} = \mu \quad \text{if} \quad y_i > 0\,; $$
$$\leq \mu \quad \text{if} \quad y_i = 0\,. \tag{13}$$

Since $X > 0$, some $x_i > 0$, so that $\mu > 0$. Hence we deduce from (13) that if $y_i > 0$ then $x_i > 0$. Thus if $x_i = 0$ we have $v_i \beta_i \leq \lambda$ from (11). If now $x_i > 0$ equality holds in (11) and the quantity $\beta_i x_i e^{-\alpha_i y_i} > 0$; hence $v_i \beta_i > \lambda$. Thus for the "inside" problem we have deduced the criterion that there is a λ such that:

$$x_i > 0 \quad \textit{if and only if} \quad v_i \beta_i > \lambda\,. \tag{14}$$

Now suppose $y_i > 0$. Then $x_i > 0$ and equality holds in both (11) and (13). We divide (11) into (13) and get $\lambda \alpha_i x_i = \mu$. If $y_i = 0$ there are

3*

two cases. If $x_i > 0$ we get on performing the same division $\lambda \alpha_i x_i \leqq \mu$. If $x_i = 0$, obviously $0 = \lambda \alpha_i x_i < \mu$. Hence we may replace (13) by

$$\lambda \alpha_i x_i = \mu \quad \text{if} \quad y_i > 0 ;$$
$$\leqq \mu \quad \text{if} \quad y_i = 0 . \tag{15}$$

Even with this apparently simple condition we cannot find y in the general case. By combining (11) and (15) we find, as in the game case (Exercise 4) that, if $x_i > 0$ and $y_i = 0$

$$\frac{\mu \beta_i}{\lambda \alpha_i} \geq \log \frac{v_i \beta_i}{\lambda} . \tag{16}$$

But (16) is not in general a sufficient condition for y_i to be zero. It is a sufficient condition in the case $\beta_i X < 1$, and that is why the game problem that results can be solved. The author does not know how to find y_i in the general case (Exercise 6). From (15) one can deduce that $x_i \alpha_i \geqq x_j \alpha_j$ for all j if $y_i > 0$. However the main problem here, to determine the solution for the "outside" or y-player, remains unsolved.

We turn to the problem $\underset{x}{\text{Max}} \underset{y}{\text{Min}} F(x, y)$. In this case, there is for fixed x no guarantee (unless $\beta_i X < 1$) from convexity considerations that the y-solution will be unique. Thus writing, as usual,

$$\varphi(x) = \underset{y}{\text{Min}} F(x, y) ,$$

we have to use the general formula (12) of Theorem I of Chapter III:

$$D_y \varphi(x) = \underset{y \in Y(x)}{\text{Min}} \Sigma \gamma_i v_i \beta_i e^{-\alpha_i y_i} e^{-\beta_i x_i e^{-\alpha_i y_i}} . \tag{17}$$

Each solution $y \in Y(x)$ will satisfy a condition in appearance of the form (13), this time as a consequence of its solving an elementary minimum problem. The μ will of course depend on y.

We now apply our version of the Lagrange multiplier theorem, Theorem VII of Chapter III. If x is the solution there exists a λ such that $D_y \varphi(x) \leqq \lambda \Sigma \gamma_i$ for any possible direction γ. Suppose $x_i > 0$. Then the direction with $\gamma_i = -1$ and $\gamma_j = 0$ for $j \neq i$ is possible, and for that γ

$$D_y \varphi(x) \leqq -\lambda . \tag{18}$$

Using formula (17) we get

$$\underset{y \in Y(x)}{\text{Min}} \left[-v_i \beta_i e^{-\alpha_i y_i} e^{-\beta_i x_i e^{-\alpha_i y_i}} \right] \leqq -\lambda ,$$

or

$$\underset{y \in Y(x)}{\text{Max}} \left[v_i \beta_i e^{-\alpha_i y_i} e^{-\beta_i x_i e^{-\alpha_i y_i}} \right] \geqq \lambda . \tag{19}$$

It follows trivially from (19) that, if $x_i > 0$, $v_i \beta_i > \lambda$.

As we said above, conditions (13) hold for each solution $y \in Y(x)$. As in the Min-Max problem we see that $\mu > 0$, so that $y_i > 0$ only if

$x_i > 0$. Suppose then $x_i = 0$, and differentiate in the direction γ' with $\gamma'_i = +1$ and all other $\gamma'_j = 0$. We have from our Lagrange theorem that

$$D_{\gamma'}\varphi(x) \leqq \lambda \tag{19}$$

and from formula (17) that

$$D_{\gamma'}\varphi(x) = v_i\beta_i. \tag{20}$$

Thus if $x_i = 0$ we have $v_i\beta_i \leqq \lambda$. Hence, combining with the result of the previous paragraph:

$$x_i > 0 \quad \textit{if and only if} \quad v_i\beta_i > \lambda. \tag{21}$$

As to y, the matter is, as in the Min-Max problem, more difficult; because of the non-unique nature of y given x it is more difficult than in the Min-Max problem. However we may consider this problem more satisfactorily resolved, because we have this time obtained the criterion for the "outside" player, i.e. (21).

The reader will appreciate as has been noted before that (14) and (21) are not, except when Max-Min is equal to Min-Max, the same thing.

We shall now work out the Max-Min and Min-Max problems for (8) in a particular case:

$$n = 2, \quad v_1 = v_2 = \alpha_1 = \alpha_2 = X = Y = 1, \quad \beta_1 = \beta_2 = 10. \tag{22}$$

We begin with the Min-Max case. Condition (11) becomes

$$10e^{-y_i}e^{-10x_ie^{-y_i}} = \lambda \quad \text{if} \quad x_i > 0;$$
$$\leqq \lambda \quad \text{if} \quad x_i = 0, \tag{23}$$

and condition (15) becomes

$$\lambda x_i = \mu \quad \text{if} \quad y_i > 0;$$
$$\leqq \mu \quad \text{if} \quad y_i = 0. \tag{24}$$

The criterion (14) reduces to $10 > \lambda$; hence both x_1 and x_2 are positive. We may suppose without loss of generality that $y_1 \geqq y_2$. Then $y_1 > 0$, so that, using (24), $x_1 \geqq x_2$. Now suppose $y_2 > 0$. Then, from (24), $x_1 = x_2 = 1/2$. In doing Exercise 3 the reader will have noted that $e^{-\alpha y}e^{-\beta xe^{-\alpha y}}$ is increasing in y if $\beta xe^{-\alpha y} > 1$. In the case at hand $\beta_i x_i e^{-\alpha_i y_i} = 5e^{-y_i} > 5e^{-1} > 1$, so that indeed the left side of (23) is a strictly increasing function of y_i. Since x_1 and x_2 are positive, equality holds in (23); since $x_1 = x_2$ and because the left side is increasing, the solutions for y_1 and y_2 are unique and the same: $y_1 = y_2$.

What we have just proved in the Min-Max problem is that if $y_1 \geqq y_2$ the solutions are either $y = (1, 0)$ and $x = (x_1, x_2)$ with $x_1 \geqq x_2$ and determined from (23), or else $y = (1/2, 1/2)$ and $x = (1/2, 1/2)$. There is of course another possibility, $y = (0, 1)$; but this is the mirror of $(1, 0)$.

We check first the possibility $y = (1, 0)$. Using (23) we find (Exercise 7) that

$$x_1 = \frac{9e}{10(e+1)} \doteq .658; \quad x_2 = \frac{10+e}{10(e+1)} \doteq .342;$$

$$\lambda = 10 \exp\left[-\frac{10+e}{e+1}\right] \doteq .32697 \tag{25}$$

so that

$$F(x, y) = 1 - e^{-10x_1 e^{-y_1}} + 1 - e^{-10x_2 e^{-y_2}} = 1 - \frac{\lambda e}{10} + 1 - \frac{\lambda}{10}$$

$$= 2 - \frac{\lambda(e+1)}{10} \doteq 1.88 . \tag{26}$$

The result for the possibility $y = (1/2, 1/2)$ is

$$F(x, y) = 2(1 - e^{-5e^{-.5}}) \doteq 1.904 . \tag{27}$$

Thus comparing (27) with (26) we see that the answer is either $y = (1, 0)$ or $y = (0, 1)$. In the first case x_1 and x_2 are given by (25); in the second case x_1 and x_2 are reversed in (25).

We turn now to the Max-Min case, under the assumptions (22). Using the criterion (21), we see that both x_1 and x_2 are positive. Now differentiate in the direction

$$\gamma = (-1/\sqrt{2}, 1/\sqrt{2}) .$$

We must have

$$D_\gamma \varphi(x) \leqq 0 . \tag{28}$$

Using formula (17), we get

$$D_\gamma \varphi(x) = -\frac{10}{\sqrt{2}} e^{-y_1} e^{-10x_1 e^{-y_1}} + \frac{10}{\sqrt{2}} e^{-y_2} e^{-10x_2 e^{-y_2}} \tag{29}$$

where $y = (y_1, y_2)$ is some element of $Y(x)$. Since it minimizes against x, the y must for some μ satisfy the condition

$$10x_i e^{-y_i} e^{-10x_i e^{-y_i}} = \mu \quad \text{if} \quad y_i > 0 ;$$

$$\leqq \mu \quad \text{if} \quad y_i = 0 . \tag{30}$$

Suppose first that $y_2 > 0$. Then from (30) we have

$$x_1 e^{-y_1} e^{-10x_1 e^{-y_1}} \leqq x_2 e^{-y_2} e^{-10x_2 e^{-y_2}} . \tag{31}$$

Combining (28) and (29), we get

$$e^{-y_1} e^{-10x_1 e^{-y_1}} \geqq e^{-y_2} e^{-10x_2 e^{-y_2}} . \tag{32}$$

Divide (32) into (31); we get $x_2 \geqq x_1$. Now suppose $y_2 = 0$. Then $y_1 = 1$ and (32) reads

$$e^{-10x_2} \leqq e^{-1} e^{-10x_1 e^{-1}},\tag{33}$$

so that

$$x_1 \leqq \frac{9e}{10(e+1)} \doteq .658.\tag{34}$$

Hence, whether $y_2 > 0$ or $y_2 = 0$, (34) holds. We have the same result for x_2, so that finally

$$.342 \doteq \frac{10+e}{10(e+1)} \leqq x_1, x_2 \leqq \frac{9e}{10(e+1)} \doteq .658.\tag{35}$$

Now if the x_i satisfy (35)

$$\beta_i x_i e^{-\alpha_i y_i} = 10 x_i e^{-y_i} \geqq \frac{10+e}{e+1} e^{-1} > 1,$$

so that (Exercise 3) $F(x, y)$ is concave in y for each such fixed x. Hence for x satisfying (35) the solution for y is on a corner (Exercise 6, Chapter III). Hence we plot the two curves C_1 and C_2 as follows: C_1 has $y = (1, 0)$ and C_2 has $y = (0, 1)$. We plot these curves in terms of x_1 for all values of x_1, x_2 satisfying $0 \leqq x_1, x_2 \leqq 1$, in Fig. 4. The curves are given by

$$C_1:\ 1 - e^{-10x_1 e^{-1}} + 1 - e^{-10x_2};$$
$$C_2:\ 1 - e^{-10x_1} + 1 - e^{-10x_2 e^{-1}}.$$

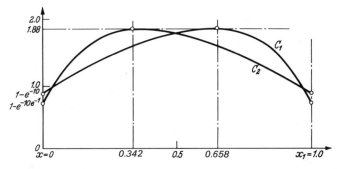

Fig. 4. The payoff for the x-player against the strategies $y = (1, 0)$ (C_1) and $y = (0, 1)$ (C_2). Vertical scale distorted

What the solution is is now obvious from the figure. For each value of x_1, x_2 satisfying (35) the y-player will select the lower of the two curves C_1 and C_2. C_1 is strictly increasing on the interval $(.342, .5)$ and C_2 strictly decreasing on $(.5, .658)$. We already know that (35) holds for the solution. Hence the Max-Min is assumed where these curves cross, at $x_1 = x_2 = .5$. At this point

$$F(x, y) - 1 - e^{-5} + 1 - e^{-5e^{-1}} \doteq 1.834 .$$

Notice that since $1.834 < 1.88$, $\underset{x}{\text{Max}} \, \underset{y}{\text{Min}} \, F(x, y) < \underset{y}{\text{Min}} \, \underset{x}{\text{Max}} \, F(x, y)$, as asserted earlier. As to the λ for the Max-Min problem, this may be determined directly from the definition using the Dedekind-cut in Chapter III, following formula (41) there. It turns out (Exercise 8) that

$$\lambda = \frac{10e^{-5} + 10e^{-1}e^{-5e^{-1}}}{2} \doteq .32599 . \qquad (36)$$

Thus the λ's in (14) and (21) are different, if not by very much.

Note now, applying Theorem I of Chapter III, that the y which yields equality in formula (29) is the limit of the y's yielding answers to cases in which x_1 is slightly less than .5, all of which are $y = (1, 0)$. This phenomenon will be called strong backward stability. Its significance will be illustrated in the model below.

Suppose that the Max-Min problem resulted from the following model, a simple version of the general situation described near the beginning of Chapter I. There are a number of weapons systems which attack different targets. The amount (measured in proportion of budget) of the ith system is x_i, and its effectiveness β_i. The value of the ith target is v_i. The defender of the targets is allowed to strike first. He has full knowledge of the $(x_1, \ldots, x_n)$ and has adequate time to develop an optimal counter. He applies y_i units of attack to the ith system, so that the original number of weapons x_i is reduced to $x_i' = x_i e^{-\alpha_i y_i}$. This will be a "percentage vulnerable" formula in the terms of Chapters V and VII. The damage the residual quantity x_i' can do to the target is $v_i(1 - e^{-\beta_i x_i'})$. We suppose that the damage adds from target to target. Thus the total damage resulting from the choice $x = (x_1, \ldots, x_n)$ and its counter $y = (y_1, \ldots, y_n)$ is given by formula (8). Since the counter y is chosen in full knowledge of the choice of x and after that choice, we are dealing with the $\underset{x}{\text{Max}} \, \underset{y}{\text{Min}}$ problem for (8). It is very difficult in general to get

the explicit solution to this problem. The only general statement we can make is that the criterion for buying or not buying a weapons system is $v_i \beta_i$. Thus while we cannot find the exact formulas, we can at least say that if a weapons system with a given value of $v_i \beta_i$ is bought with the payoff (8) in mind, then any system with a higher value should also be bought.

As to backward stability, the numerical example computed above displays what appears at first sight to be a very curious phenomenon. Put in the terms of the above model, with two weapons systems intended for two different targets, it says that the strategy for the y-player at the optimum solution for x is to place all his attack on one system or the other; *but, if the x-player deviates from his optimum by increasing slightly the amount of the weapon of type 1 at the expense of the amount of weapon of type 2 then the y-player should attack only the weapon of type 2!* The reader is asked in Exercise 9 to explain why this is plausible. We shall define backward stability, and forward stability, strong and weak, more precisely in Chapter VII; the meaning of backward stability in the case at hand is clear.

4. A problem on allocation of defense

The problem of finding the x-strategy yielding

$$\text{Max Min } \Sigma v_i (1 - \alpha_i e^{-\kappa_i x_i/y_i})^{y_i}$$

subject to

$$\Sigma x_i = X, \quad \Sigma y_i = Y, \quad x_i \geq 0, \quad y_i > 0,$$

when $0 < \alpha_i < 1$ and the v_i and κ_i are positive, was first posed at the RAND Corporation around 1951. The function

$$f_i(x_i, y_i) = \begin{cases} v_i(1 - \alpha_i e^{-\kappa_i x_i/y_i})^{y_i} & \text{if } y_i > 0 \\ \\ v_i & \text{if } y_i = 0 \end{cases} \tag{37}$$

is intended to represent the residual value of a target if it is defended by x_i defense units and attacked by y_i attack units. It may be explained as follows: the quantity $\exp(-\kappa_i x_i/y_i)$ is the probability an individual attack unit gets through, x_i/y_i being the amount of defense against each attack unit and κ_i the effectiveness of the defense. α_i is the probability that a surviving attack unit destroys the target, the quantity $[1 - \alpha_i \exp(-\kappa_i x_i/y_i)]$ the probability the target survives the attack, and finally $[1 - \alpha_i \exp(-\kappa_i x_i/y_i)]^{y_i}$ the probability the target survives the attack with y_i weapons.

The essential character of this problem is that the x-player moves first, by constructing defenses, and that the y-player moves in full knowledge of what the x-player has done.

We have taken as the object of this section that of obtaining criteria for defending or not defending a target, in terms of the parameters. That is, we seek some quantity (or quantities) given by a formula involving the v_i, α_i, and κ_i with the property that if the quantity is large, one should defend and if it is small one should not defend.

In fact there are two formulas involving v_i, α_i, and κ_i which are crucial in the problem at hand.

According to the relative values of the quantities

$$C_i = v_i \log \frac{1}{1 - \alpha_i} \tag{38}$$

and

$$D_i = \frac{\kappa_i \alpha_i}{(1 - \alpha_i) \log \dfrac{1}{1 - \alpha_i}} \tag{39}$$

we shall be able to make partial statements as to which of the x_i are positive. In many special cases these become complete statements. For instance, if the α_i and κ_i are constants, all the targets with v_i larger than some constant are defended, i.e., $x_i^0 > 0$, and all those with v_i less than that constant are not defended. If the v_i and α_i are constants, then those with higher κ_i are defended, those with lower κ_i not.

These results appear to be the most that can be derived without a characterization of the set of solutions of the problem

$$\operatorname*{Min}_y \Sigma v_i (1 - \alpha_i e^{-\kappa_i x_i / y_i})^{y_i} \tag{40}$$

in terms of v_i, α_i, κ_i, X and Y. We have obtained all the results that appear to be possible from the Gibbs lemma, i.e., that there exists for each solution a μ such that

$$\begin{aligned} f_{iy_i}(x_i, y_i) &= \mu \quad \text{if} \quad y_i > 0, \\ &\geqq \mu \quad \text{if} \quad y_i = 0. \end{aligned} \tag{41}$$

Further results would need a detailed study of the second derivatives of $f_i(x_i, y_i)$, and are not attempted here.

The derivatives of $f_i(x_i, y_i)$. The reader (Exercise 10) will easily derive from (37) the formulas

$$\begin{aligned} f_{ix_i}(x_i, y_i) &= f_i \cdot \frac{\alpha_i \kappa_i}{\Gamma_i} e^{-\kappa_i u_i} \quad \text{if} \quad y_i > 0, \\ &= 0 \qquad\qquad\quad \text{if} \quad y_i = 0, \end{aligned} \tag{42}$$

where we have written

$$u_i = x_i / y_i$$

and

$$\Gamma_i(u_i) = 1 - \alpha_i e^{-\kappa_i u_i}. \tag{43}$$

Using the notation

$$g_i(u_i) = \log \Gamma_i(u_i) - \frac{\alpha_i \kappa_i u_i e^{-\kappa_i u_i}}{\Gamma_i(u_i)} \tag{44}$$

we find that

$$f_{iy_i}(x_i, y_i) = \begin{cases} f_i g_i(u_i) & \text{if } x_i, y_i > 0 \\ v_i(1-\alpha_i)^{y_i} \log(1-\alpha_i) & \text{if } x_i = 0, y_i \geq 0 \\ 0 & \text{if } x_i > 0, y_i = 0. \end{cases} \qquad (45)$$

There is only one difficulty in deriving f_{iy_i}, and that is in the case $x_i > 0$, $y_i = 0$. We need the well-known inequality (Exercise 11), valid for $0 < y < 1$ and $0 < r < 1$:

$$1 - r^y < y(1-r) \cdot r^{y-1}. \qquad (46)$$

We obtain

$$1 - \Gamma_i^{y_i} < y_i(1-\Gamma_i)\Gamma_i^{y_i-1} = \alpha_i y_i e^{-\kappa_i u_i} \Gamma_i^{y_i-1} < \frac{\alpha_i y_i}{1-\alpha_i} e^{-\kappa_i u_i}.$$

Hence for small y_i and $x_i > 0$

$$0 > \frac{f_i(x_i, y_i) - f_i(x_i, 0)}{y_i} > -\frac{\alpha_i v_i e^{-\kappa_i x_i/y_i}}{1-\alpha_i}.$$

It follows that as $y_i \to 0$ with x_i bounded from zero, the difference quotient approaches zero.

It results that $f_{iy_i}(x_i, y_i)$ is continuous in x_i and y_i at every point with $x_i > 0$. It is also evidently continuous at any point where $y_i > 0$. But it is discontinuous at the point $(0, 0)$, as is easily checked by taking limits along the axes (Exercise 12). Similarly $f_{x_i}(x_i, y_i)$ is two-dimensionally continuous except at the origin.

Observe that (Exercise 13)

$$\frac{d}{du_i} g(u_i) > 0 \qquad (47)$$

and

$$\frac{d}{du_i}\left[\frac{e^{-\kappa_i u_i}}{-\Gamma_i(u_i) g(u_i)}\right] < 0. \qquad (48)$$

Lack of continuity in the derivative f_{ix_i} at the origin does not destroy the validity of Theorems I and III—VII in the particular problem at hand. (This statement does not refer to BRAM's Theorem.)

We turn first to Theorem I on the existence of the directional derivative. First note that inequality (6) of Chapter III does not use the two-dimensional continuity of f_{ix_i}. However that of (11), Chapter III, does. Suppose then that $x_i = 0$ and $y_i^m \to 0$ for i in a certain set I'_0. For $i \notin I'_0$ the law of the mean may be used, the $f_{ix_i}(x_i, y_i)$ being continuous there. The remaining terms are

$$\frac{1}{d^m} \sum_{i \in I'_0} [f_i(x_i^{m'}, y_i^{m'}) - f_i(0, y_i^{m'})] \geq 0 = \sum_{i \in I'_0} \gamma_i f_{ix_i}(x_i, y_i),$$

since $f_{ix_i}(x_i, y_i^\gamma) = f_{ix_i}(0, 0) = 0$. The proof of (11) now goes through and the result of Theorem I holds (Exercise 14).

Theorem II does not apply to this problem, as the inside solution is in general multiple.

Theorem III follows from Theorem I, so that Theorem III is valid here.

As to Theorem IV, in its proof we arrive at sequences $\{x^{m*}\}$ and $\{y^{m*}\}$. Denote by I the set of those i with $x_i^0 = y_i^* = 0$. Formula (28) is replaced by (we pass from the sequence $\{m'\}$ to $\{m*\}$)

$$D_{\gamma^{m*}}\varphi(x^{m*}) = \Sigma \gamma_i^{m*} f_{ix_i}(x_i^{m*}, y_i^{m*}).$$

We may pass to the limit in the terms for which $i \notin I$ on the right; they converge to $\gamma_i^0 f_{ix_i}(x_i^0, y_i^*)$. As to the case $i \in I$, we have $\gamma_i^0 \geqq 0$ so that $\lim\limits_{m* \to \infty} \gamma_i^{m*} \geqq 0$. In addition, $f_{ix_i}(x_i, y_i)$ is bounded above on any set (Exercise 15). Hence (Exercise 16)

$$\liminf_{m* \to \infty} \gamma_i^{m*} f_{ix_i}(x_i^{m*}, y_i^{m*}) \geqq 0 = \gamma_i^0 f_{ix_i}(x_i^0, y_i^*)$$

for every $i \in I$. Combining these results we get

$$\liminf_{m* \to \infty} D_{\gamma^{m*}}\varphi(x^{m*}) \geqq \Sigma \gamma_i^0 f_{ix_i}(x_i^0, y_i^*).$$

But

$$D_{\gamma^0}\varphi(x^0) \leqq \Sigma \gamma_i^0 f_{ix_i}(x_i^0, y_i^*),$$

so that the theorem is proved for this case.

Theorems V, VI, and VII are left as exercises (17, 18, and 19) for the reader.

We do not make any explicit use of Theorems III—VII in this application.

The following lemmas are needed to prepare for Theorem I, which gives the main result in this problem.

Lemma 1. *Let* $\gamma = (1/\sqrt{n}, \ldots, 1/\sqrt{n})$, *and* $\bar{X} > X$. *There exists a constant* $a > 0$ *such that* $D_\gamma\varphi(x) \geqq a/\sqrt{n}$ *for all* x *with* $\Sigma x_i \leqq \bar{X}$.

Proof. Put $a_i = \mathrm{Min} f_{ix_i}(x, y)$ over the set defined by $0 \leqq x \leqq \bar{X}$, $y \geqq \dfrac{Y}{n}$. Put $a = \mathrm{Min}\limits_i a_i$. Then

$$D_\gamma\varphi(x) = \mathrm{Min}_{y \in Y(x)} \Sigma \frac{1}{\sqrt{n}} f_{ix_i}(x_i, y_i) \geqq (a/\sqrt{n}),$$

since at least one $y_i \geqq Y/n$.

Lemma 2. *Suppose that* $x_i > 0$, $x^m \to x$, *and* $y^m \in Y(x^m)$. *Suppose also that* $y_i^m > 0$ *for all* m. *Then it is impossible that* $y_i^m \to 0$.

Proof. Suppose that $y_i^m \to 0$; one may suppose that the sequence $\{y^m\}$

converges to a limit y. We then have that there exist $\mu, \mu_1, \ldots, \mu_m, \ldots$ such that for all j

$$f_{jy_j}(x_j, y_j) = \mu \quad \text{if} \quad y_j > 0;$$

$$\geqq \mu \quad \text{if} \quad y_j = 0, \tag{49}$$

and for $m = 1, 2, \ldots,$

$$f_{jy_j}(x_j^m, y_j^m) = \mu_m \quad \text{if} \quad y_j^m > 0;$$

$$\geqq \mu_m \quad \text{if} \quad y_j^m = 0. \tag{50}$$

Since $f_{jy_j}(x_j, y_j) < 0$ except when $y_j = 0$, μ and all the μ_m are negative. Since $\Sigma y_j = Y$, there exists a k such that $y_k > 0$. f_{ky_k} is uniformly continuous in the neighborhood of the point (x_k, y_k) so that

$$\lim_{m \to \infty} \mu_m = \lim_{m \to \infty} f_{ky_k}(x_k^m, y_k^m) = f_{ky_k}(x_k, y_k) = \mu. \tag{51}$$

But also

$$\lim_{m \to \infty} \mu_m = \lim_{m \to \infty} f_{iy_i}(x_i^m, y_i^m) = f_{iy_i}(x_i, 0) = 0. \tag{52}$$

Since $\mu < 0$, (51) and (52) are in contradiction. Thus the lemma is proved.

Lemma 3. If x^0 maximizes $\varphi(x)$ and $x_i^0 > 0$, then there is a $y^0 \in Y(x^0)$ such that $y_i^0 > 0$.

Proof. Suppose that the sequence $x^m \to x^0$ is as follows: $x_i^m \to x_i^0$ from the left, i.e. $x_i^m < x_i^0$, and $x_j^m = x_j^0$ for all $j \neq i$. Let $y^m \in Y(x^m)$. Suppose the sequence $m \to \infty$ chosen also so that $\{y^m\}$ converges to a limit y^0. Suppose first that all the y_i^m after a certain point, say for $m \geqq m_0$, are zero. Evidently $y^0 \in Y(x^0)$ and

$$\varphi(x^{m_0}) = v_i + \sum_{j \neq i} f_j(x_j^{m_0}, y_j^{m_0}) = v_i + \sum_{j \neq i} f_j(x_j^0, y_j^{m_0})$$

$$\geqq v_i + \sum_{j \neq i} f_j(x_j^0, y_j^0) = \varphi(x^0).$$

Since obviously $\varphi(x^{m_0}) \leqq \varphi(x^0)$, then $\varphi(x^{m_0}) = \varphi(x^0)$. Put

$$\sigma = x_i^0 - x_i^{m_0}.$$

Then $\sigma > 0$. Put

$$x_i^\sigma = x_i^{m_0} + \sigma/n$$

$$x_j^\sigma = x_j^0 + \sigma/n \qquad j \neq i.$$

From the Law of the Mean (Theorem VI of Chapter III) we now have, for some point x on the segment joining x^{m_0} to x^σ,

$$\frac{\varphi(x^\sigma) - \varphi(x^{m_0})}{|x^\sigma - x^{m_0}|} \geqq D_\gamma \varphi(x). \tag{53}$$

Using now Lemma 1, we get

$$\varphi(x^\sigma) - \varphi(x^{m_0}) \geq |x^\sigma - x^{m_0}| \cdot \frac{a}{\sqrt{n}} = \frac{\sigma}{n} \cdot \sqrt{n} \cdot \frac{a}{\sqrt{n}} = \frac{a\sigma}{n}.$$

However evidently

$$\Sigma x_i^\sigma = \Sigma x_i^0 = X.$$

Hence x^0 does not yield the maximum.

It follows that $y_i^m > 0$ for infinitely many m. We pass to the corresponding subsequence. Lemma 2 now applies. It follows that $y_i^0 > 0$, and Lemma 3 is proved.

Definition. We denote by $Y^\gamma(x)$ the collection of those elements of $Y(x)$ which yield the directional derivative in the direction γ and each of which is the limit of a sequence of points $y^m \in Y(x^m)$, with $x^m \neq x^0$ and $x^m \to x^0$ along the vector of direction γ issuing from x^0.

Lemma 4. *Suppose that $x^0 = (x_1^0, ..., x_n^0)$ maximizes $\varphi(x)$ subject to $\Sigma x_i = X$, $x_i \geq 0$, and let γ be an admissible direction for x^0. Let I^- be the set of those i's for which $\gamma_i < 0$. Then there is a $y^\gamma \in Y^\gamma$ and an $i \in I^-$ such that $y_i^\gamma > 0$.*

Proof. We denote by I^+ the set of i's with $\gamma_i > 0$, and by I^0 the set with $\gamma_i = 0$.

Let $x^m \to x^0$ from the direction γ, with $x^m \neq x^0$. Let $y^m \in Y(x^m)$, and suppose the sequence $\{x^m\}$ so chosen that the y^m converge. Then the y^m converge, as follows from the proof of Theorem I of Chapter III and its extension to this case, to an element $y^\gamma \in Y(x^0)$, and we have

$$D_\gamma \varphi(x^0) = \Sigma \gamma_i f_{ix_i}(x_i, y_i^\gamma).$$

Suppose that $y_i^\gamma = 0$ for all $i \in I^-$. We may now write

$$D_\gamma \varphi(x^0) = \sum_{i \in I^+} \gamma_i f_{ix_i}(x_i, y_i).$$

Since γ is admissible and x^0 yields the maximum, $D_\gamma \varphi(x^0) \leq 0$. Hence $y_i^\gamma = 0$ for all $i \in I^+$ as well. Hence $y_i^m \to 0$ for all $i \in I^- \cup I^+$.

Suppose that $i \in I^-$, or that $i \in I^+$ and $x_i^0 > 0$. In either case $x_i^0 > 0$, and $y_i^m \to 0$. Lemma 2 now implies that $y_i^m = 0$ for m sufficiently large. Take m_0 such that if $i \in I^-$, or $i \in I^+$ and $x_i^0 > 0$, $y_i^m = 0$ for $m \geq m_0$. We also take m_0 sufficiently large that $x_i^m > 0$ for $i \in I^-$.

Fix on $m \geq m_0$. We have

$$\varphi(x^m) = \sum_{i \in I^-} v_i + \sum_{\substack{i \in I^+ \\ x_i^0 > 0}} v_i + \sum_{\substack{i \in I^+ \\ x_i^0 = 0}} f_i(x_i^m, y_i^m) + \sum_{i \in I^0} f_i(x_i^m, y_i^m)$$

$$\geq \sum_{i \in I^-} v_i + \sum_{\substack{i \in I^+ \\ x_i^0 > 0}} v_i + \sum_{\substack{i \in I^+ \\ x_i^0 = 0}} f_i(0, y_i^m) + \sum_{i \in I^0} f_i(x_i^0, y_i^m),$$

since $f_i(x_i, y_i)$ is non-decreasing in x for fixed y and since $x_i^m = x_i^0$ for $i \in I^0$. Now

$$\sum_{\substack{i \in I^+ \\ x_i^0 = 0}} y_i^m + \sum_{i \in I^0} y_i^m = Y \, ;$$

and y^γ, which has $y_i^\gamma = 0$ for $i \in I^- \cup I^+$, minimizes against x^0 among vectors with $y_i \geq 0$, $\Sigma y_i = Y$. It follows that

$$\varphi(x^m) \geq \sum_{i \in I^-} v_i + \sum_{\substack{i \in I^+ \\ x_i^0 > 0}} v_i + \sum_{\substack{i \in I^+ \\ x_i^0 = 0}} f_i(0, y_i^\gamma) + \sum_{i \in I^0} f_i(x_i^0, y_i^\gamma) = \varphi(x^0) \, .$$

Thus x^m also yields the maximum. Let $i \in I^-$. We have $x_i^m > 0$ for that i, and thus may apply Lemma 3: $\bar{y}_i^m > 0$ for some $\bar{y}^m \in Y(x^m)$. Choose a subsequence $\{m'\}$ so that $\bar{y}^{m'}$ converges to a limit $\bar{y}^\gamma$. Evidently $\bar{y}^\gamma \in Y^\gamma(x^0)$. Since $\bar{y}_i^{m'} > 0$ for each m', then Lemma 2 implies that $\bar{y}_i^\gamma > 0$.

Thus the supposition that $y_i^\gamma = 0$ for all $i \in I^-$ leads to the conclusion that there is for each $i \in I^-$ a $\bar{y}^\gamma$ (depending on i) with $\bar{y}_i^\gamma > 0$. The lemma is thus proved.

The main result for this problem is the following.

Theorem I. Let C_i and D_i be defined by (38) and (39) above. Suppose $x_i^0 > 0$. Then

(a) if $C_j \geq C_i$ and $D_j \geq D_i$, $x_j^0 > 0$;

(b) if $C_j \geq C_i$ and $D_j < D_i$, either $x_j^0 > 0$, or $x_j^0 = 0$ and $y_j > 0$ for some $y \in Y(x^0)$ which also has $y_i > 0$;

(c) if $C_j < C_i$ and $D_j \geq D_i$ either $x_j^0 > 0$, or $x_j^0 = 0$ and $y_j = 0$ for some $y \in Y(x^0)$ which has $y_i > 0$.

Proof. Consider the direction γ defined by: $\gamma_i = -1/\sqrt{2}$, $\gamma_j = 1/\sqrt{2}$, $\gamma_k = 0$ for $k \neq i, j$. γ is admissible, so that

$$D_\gamma \varphi(x^0) \leq 0 \, . \tag{54}$$

From Lemma 4 there is a $y^\gamma \in Y(x^0)$ with $y_i^\gamma > 0$. We shall fix on this y^γ throughout. The theorem results from the two statements following:

(i) if $y_j^\gamma = 0$ and $x_j^0 = 0$, $C_j < C_i$;

(ii) if $y_j^\gamma > 0$ and $x_j^0 = 0$, $D_j < D_i$.

Proof of statement (i). If $y_j^\gamma = 0$, we get

$$f_{iy_i}(x_i^0, y_i^\gamma) = \mu \leq f_{jy_j}(x_j^0, y_j^\gamma) \, . \tag{55}$$

Since $x_j^0 = 0$ we may apply the middle line of (45) to the right side of (55) and the top line of (45) to the left side of (55). We get

$$f_i(x_i, y_i^\gamma) \, g(u_i) \leq v_j (1 - \alpha_j)^{y_j^\gamma} \log(1 - \alpha_j) \, . \tag{56}$$

Hence, using (47) on the left and putting $y_j^\gamma = 0$ on the right,

$$f_i(x_i, y_i^\gamma) \log \frac{1}{1 - \alpha_i} > v_j \log \frac{1}{1 - \alpha_j} \, . \tag{57}$$

Since $y_i^\gamma > 0$, $f_i(x_i, y_i^\gamma) < v_i$, and we may replace (57) by

$$v_i \log \frac{1}{1-\alpha_i} > v_j \log \frac{1}{1-\alpha_j}, \tag{58}$$

i.e. $C_i > C_j$. Statement (i) is proved.

We turn to statement (ii). Suppose $y_j^\gamma > 0$ and $x_j^0 = 0$. We now have

$$f_{iy_i}(x_i^0, y_i^\gamma) = \mu = f_{jy_j}(x_j^0, y_j^\gamma),$$

so that, using (45),

$$-f_i g(u_i) = -f_j g(u_j) \tag{59}$$

with both sides of (59) positive. Now using (54) and Theorem I of Chapter III as extended to the present case, we get

$$\gamma_i f_{ix_i}(x_i^0, y_i^\gamma) + \gamma_j f_{jx_j}(x_j^0, y_j^\gamma) \leq 0,$$

i.e.

$$f_{jx_j}(x_j^0, y_j^\gamma) \leq f_{ix_i}(x_i^0, y_i^\gamma). \tag{60}$$

Since $y_i^\gamma > 0$ and $y_j^\gamma > 0$ we may use the top line of (42) and get

$$\frac{f_i \alpha_i \kappa_i}{\Gamma_i} e^{-\kappa_i u_i} \geq \frac{f_j \alpha_j \kappa_j}{\Gamma_j} \tag{61}$$

On dividing (61) by (59) we get

$$\frac{\alpha_i \kappa_i e^{-\kappa_i u_i}}{-g_i \Gamma_i} \geq \frac{\alpha_j \kappa_j}{-g_j \Gamma_j}. \tag{62}$$

Since $u_j = 0$ we may replace (62) by

$$\frac{\alpha_i \kappa_i e^{-\kappa_i u_i}}{-g_i \Gamma_i} \geq \frac{\alpha_j \kappa_j}{(1-\alpha_j) \log \dfrac{1}{1-\alpha_j}}. \tag{63}$$

Finally, because of (48) we may replace the left side of (63) by a larger expression:

$$\frac{\alpha_i \kappa_i}{(1-\alpha_i) \log \dfrac{1}{1-\alpha_i}} > \frac{\alpha_j \kappa_j}{(1-\alpha_j) \log \dfrac{1}{1-\alpha_j}}. \tag{64}$$

Thus $D_i > D_j$ and the statement (ii) is proved.

This proves Theorem I.

Interpretation. Theorem I may be expressed in words as follows: Suppose the ith target is defended in an optimal strategy. Then

(a) if both $C_j \geq C_i$ and $D_j \geq D_i$, the jth target will also be defended;

(b) if $C_j \geq C_i$ and $D_j < D_i$, either the jth target is defended or else it is not, in which case there is an attack against the optimal defense which hits both;

(c) if $C_j < C_i$ and $D_j \geqq D_i$, either the jth target is defended or it is not, in which case there is an attack against the optimal defense which hits the ith but not the jth target.

The case when the C_i and D_i are ordered in the same way is particularly easy. Then all the $x_i^0 > 0$ from some point on in the ordering. Examples are:

(a) v_i variable but α and κ are constant[7].

(b) κ_i variable but α and v constant.

A geometrical version of Theorem I may be convenient. One plots, from left to right on an abscissa, increasing values of D_i. From bottom to top one plots the increasing values of C_i. To each i there corresponds a point on the resulting display. There will be n points corresponding to targets on the resulting graph as in Fig. 5. The heavy dots represent targets. The theorem states that if $x_i^0 > 0$ for a target, all the $x_j^0 > 0$ for targets to the right or above i (both in the sense $\geqq$).

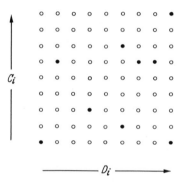

Fig. 5. Geometrical interpretation of Theorem I

For a target up ($\geqq$) to the left ($<$), if $x_j^0 = 0$ there is an attack against the optimal defense strategy in which both the ith target and the jth target are attacked. For targets down ($<$) to the right ($\geqq$) either $x_j^0 > 0$ or there is an attack against the optimal defense strategy in which the ith target but not the jth target is attacked.

5. Conclusion

The foregoing examples have illustrated the basic technique of dealing with the directional derivative. As in simple maximum problems, it is often impossible to obtain complete solutions without careful study of the second derivative. The entirety of Chapter V deals with a problem in which the second derivative information was the most useful.

---

[7] This is the way it was first posed to this writer.

As to the examples in this chapter, it is now clear that the Lagrange multiplier principle is not always useful in Max-Min problems. For instance in the examples of Sections 3 and 4, we found it more convenient to differentiate in 45° directions directly. We had to know that the y_i corresponding to the down side $(\gamma_i = -1/\sqrt{2})$ and the y_j corresponding to the up side $(\gamma_j = 1/\sqrt{2})$ were from the same solution $y^\gamma \in Y(x^0)$. Without this we could not have proved that $x_2 \geqq x_1$ by dividing (32) into (31) in Section 3, nor could we have written inequality (55) using the same μ, for in a general problem two $y \in Y(x^0)$ may well have different μ's.

We discussed *backward stability* in the numerical example of Section 3. Another example of this phenomenon is represented by Lemma 4 in Section 4. A formal definition of this is given in Chapter VII. If the outside player, as in the example of Section 4, is thought of as a defender, one might interpret it in less elegant terms as follows: "If you lower your guard, you may be hit." But the phenomenon appears, as in the example of Section 3, in places where such an easy explanation does not work. In the model mentioned in Section 3, it would mean "If you buy more forces of type 1 at the expense of type 2, the best your opponent can do is to forget about type 1 and go to type 2." The problem of forward and backward stability is central to the analysis of Max-Min problems, and deserves further study.

Exercises to Chapter IV

(The author does not know the answer to the problem marked with an asterisk.)

1. Work out the problem $\underset{y}{\text{Min}} \underset{x}{\text{Max}} F(x, y)$ for Section 1 [formula (1)]. Do it directly and by use of the theory of Chapter III, and show that $\underset{x}{\text{Max}} \underset{y}{\text{Min}} F(x, y) < \underset{y}{\text{Min}} \underset{x}{\text{Max}} F(x, y)$.

2. Show that given x with $\Sigma x_i > 0$ the solution to the problem $\underset{y}{\text{Min}} F(x, y)$ defined by formula (5) and with $\Sigma y_i = Y$, $y_i \geqq 0$, is unique.

3. Verify that the function $g_i(y_i) = 1 - e^{-\beta_i x_i e^{-\alpha_i y_i}}$ is concave or convex in y depending on whether $\beta_i x_i e^{-\alpha_i y_i} > 1$ or $\beta_i x_i e^{-\alpha_i y_i} < 1$.

4. Solve the problem (8) in the case that $\beta_i X < 1$, $i = 1, \ldots, n$ by finite allocation game techniques. Observe that y_i need not be positive when x_i is, and that in fact a necessary and sufficient condition that $y_i > 0$, if $x_i > 0$, is that

$$\frac{\mu \beta_i}{\lambda \alpha_i} < \log \frac{v_i \beta_i}{\lambda},$$

where λ and μ are the Lagrange multipliers associated with the x- and y-players respectively.

5. We continue with problem (8) regarded as a Min-Max problem. In the case $\beta_i X > 1$ the condition mentioned in Exercise (4) is not always sufficient to imply $y_i = 0$. Explain.

6.* Attempt to determine the sufficient conditions, for the Min-Max problem (8), under which $y_i > 0$, or alternatively results analogous to Theorem I in Section 4.

7. Carry out the solution (25).

8. Derive formula (36) from the Dedekind-cut definition of λ in Chapter III. For which up vector γ is the maximum in (41), Chapter III, attained in the case at hand?

9. Explain why backward stability in the example of the end of Section 3 is plausible. That is, why should, in the Max-Min problem under condition (37), the y-player attack only the *second* system if the x-player deviates from his optimum by increasing the amount of the first and decreasing the amount of the second?

10. Calculate the derivatives (42), with special care when $y_i = 0$.

11. Use the (elementary) law of the mean to prove the inequality (46).

12. Verify the statements in the text just above formula (47) concerning limits along the axes of $f_{x_i}(x_i, y_i)$ and $f_{y_i}(x_i, y_i)$. Take, as well, limits from arbitrary directions θ, $0 \leq \theta \leq \pi/2$.

13. Verify formulas (47) and (48).

14. Write out in full the proof of Theorem I for the problem of Section 4.

15. Prove that the $f_{ix_i}(x, y)$ as given by (42) is bounded over the whole space.

16. Prove that if $a_n \to 0$ and $0 \leq b_n \leq b$ for all n then $\liminf_{n \to \infty} a_n b_n \geq 0$.

17. Prove Theorem V for the case of Section 4.

18. Prove Theorem VI for the case of Section 4.

19. Prove Theorem VII for the case of Section 4.

Chapter V

A basic weapons selection model

The particular problem treated in this chapter was first posed[8] for the purpose of studying the theoretical differences between "percentage vulnerable" systems, in which the action of the counter y on x was of the form $xe^{-\alpha y}$, and "numerically vulnerable" systems, in which this action was represented by $xe^{-\alpha y/x}$. A typical example of the first is search for submarines, in which the number of undetected submarines is equal to the number present times the probability that each submarine is not detected. With twice as many submarines and the same search effort, the number of detections (and survivals from detections) is twice as great. A typical example of the second is an attack on Minuteman silos, in which the countering effort y must be parcelled out in amounts y/x to each of the silos. The quantity $xe^{-\alpha y/x}$ then gives the number of surviving silos. In this case, doubling the number of silos will not double the number of kills with a given counter-effort: it will increase it, but in a more complicated way. What was needed was a theoretical basis for selecting among these systems, taking account of their military values and vulnerabilities α.

The basic outline is as follows. The problem is stated in formulas (1)—(4) below. We first compute the directional derivative of $\varphi(x)$ for the "second special case," with only numerically vulnerable systems present, the "first special case" (percentage vulnerable only) being a problem studied already in Chapter II. This computation is necessary because the ratio y_j/x_j appears and we need derivatives when $x_j = 0$; thus the hypotheses of the fundamental Theorem I of Chapter III fail. This derivative turns out to be a continuous function. It is not however itself differentiable everywhere either in the ordinary sense or the directional derivative sense. We compute therefore its difference quotients and prove from them that $\varphi(x)$ is convex. This fact proves that corner solutions will do. We then analyze the possibility of non-extreme solutions, and find them (Theorem I) trivial.

We then go through the same process for the general problem, with terms of both types present. This is much complicated by the appearance and disappearance of terms in the formulas, a difficulty surmounted by passing to a space with one more dimension [the (x, μ)-space used

---

[8] By Dr. THOMAS PHIPPS.

in both the second special problem, where it was not essential, and in the general problem, where it was]. We again calculate the difference quotients, which again proves (Theorem III) that except for trivialities, only one system of numerically-vulnerable type (the index used is j) can enter in the mix.

The problem in the general case is thus reduced to studying mixes with one numerically-vulnerable system and possibly several percentage-vulnerable systems (indexed by i). The method of obtaining the solution is the main object of section 3, which presents detailed information on a function $H(\xi)$, with the aid of which it is possible to solve for the values of the x_i and of the single x_j by hand or machine. In some instances a study of this function makes it possible to read off results directly from the coefficients. Such results are given in Theorems IV and V.

The Addendum considers the problem arising when some countering action, for instance antisubmarine search, affects several percentage-vulnerable systems simultaneously. The result is that only one representative of each simultaneously vulnerable group will enter into the optimal mix. Thus the problem is reduced to the basic problem.

The chapter concludes with two numerical examples and an example proving that the problem is not a game.

This model is linear in the sense that residual weapons of type i, j have unit values v_i, v_j. There is an important instance in which a Max-Min-Max non-linear model decomposes into a linear model of the present type and an allocation problem. This is discussed in Chapter VII.

The problem

Put

$$F(x, y) = \sum_{i=1}^{n} v_i x_i e^{-\alpha_i y_i} + \sum_{j=n+1}^{m} v_j x_j e^{-\alpha_j y_j / x_j}, \tag{1}$$

and, as usual,

$$\varphi(x) = \underset{y}{\text{Min}}\, F(x, y), \tag{2}$$

where

$$\Sigma y_i + \Sigma y_j = 1, \quad y_i, y_j \geq 0. \tag{3}$$

The problem is to find the x-strategy yielding

$$\underset{x}{\text{Max}}\, \varphi(x) = \underset{x}{\text{Max}} \underset{y}{\text{Min}}\, F(x, y)$$

subject to the restrictions

$$\Sigma x_i + \Sigma x_j = 1, \quad x_i, x_j \geq 0. \tag{4}$$

We cannot apply the theory of Chapter III directly. The first partial derivatives of $F(x, y)$ are not continuous at points where $x_j = y_j = 0$ for some $j = n + 1, \ldots, m$. But we may nevertheless still calculate the derivatives of $\varphi(x)$.

The problem cannot be separated into two problems corresponding to the sets of variables indexed by i and j, since the side conditions (1) and (2) permit shifts of resources from one "problem" to the other.

Nevertheless it is useful to study these two problems separately. The first special problem,

$$\text{Max}_x \text{ Min}_y \sum_{i=1}^{n} v_i x_i e^{-\alpha_i y_i} \tag{5}$$

subject to

$$\Sigma x_i = \Sigma y_i = 1, \quad x_i \geq 0, \quad y_i \geq 0,$$

has long since been solved, and is completely treated in Chapter II.

1. The second special problem. The more difficult of these problems is

$$\text{Max}_x \text{ Min}_y H(x, y),$$

where

$$H(x, y) = \Sigma_j v_j x_j e^{-\alpha_j y_j / x_j}. \tag{6}$$

The function

$$f(x, y) = x e^{-\alpha y / x}$$

is taken to be zero if $x = 0$. It is easy to calculate

$$f_x(x, y) = \left(1 + \frac{\alpha y}{x}\right) e^{-\alpha y / x} \quad \text{if} \quad x > 0;$$

$$f_x(0, y) = 0 \qquad\qquad \text{if} \quad y > 0;$$

$$f_x(0, 0) = 1.$$

This derivative is obviously discontinuous at $x = y = 0$.

We need the solution for y in terms of x yielding $\varphi(x) = \text{Min}_y H(x, y)$. We suppose $\Sigma x_j > 0$. We have for some μ

$$\alpha_j v_j e^{-\alpha_j y_j / x_j} = \mu \quad \text{if} \quad x_j > 0 \quad \text{and} \quad y_j > 0; \tag{7}$$

$$\alpha_j v_j e^{-\alpha_j y_j / x_j} \leq \mu \quad \text{if} \quad x_j > 0 \quad \text{and} \quad y_j = 0; \tag{8}$$

$$0 = \mu \quad \text{if} \quad x_j = 0 \quad \text{and} \quad y_j > 0; \tag{9}$$

$$0 \leq \mu \quad \text{if} \quad x_j = 0 \quad \text{and} \quad y_j = 0. \tag{10}$$

Since some $x_j > 0$, we have from (7) and (8) that $\mu > 0$. From (9) it follows that $x_j = 0$ implies $y_j = 0$. Suppose $\alpha_j v_j \leq \mu$. If $y_j > 0$, then $x_j > 0$ and we have a contradiction from (7). Hence $y_j = 0$. Suppose

$\alpha_j v_j > \mu$ and $x_j > 0$. Then $y_j > 0$ from (8). Hence from (7) $y_j = \dfrac{x_j}{\alpha_j} \log \dfrac{\alpha_j v_j}{\mu}$.
Hence the solution may be characterized as follows:

$$y_j = \frac{x_j}{\alpha_j} \log \frac{\alpha_j v_j}{\mu} \quad \text{if} \quad x_j > 0 \quad \text{and} \quad \alpha_j v_j > \mu;$$

$$y_j = 0 \qquad \qquad \text{otherwise}.$$ (11)

μ is determined uniquely from the equation

$$\sum_{\alpha_j v_j > \mu} \frac{x_j}{\alpha_j} \log \frac{\alpha_j v_j}{\mu} = 1.$$ (12)

Since the left side of (12) is decreasing strictly in μ and goes from large
values for small μ to zero if $\mu \geq \alpha_j v_j$ for all j, then (12) has a unique
solution for μ. For values of μ not equal to any of the $\alpha_j v_j$, μ is continuous
and continuously differentiable in x, as is easily seen by solving (12) for
$\log \mu$. Suppose $\mu^0 = \alpha_j v_j$ for some value x^0. Suppose $x^n \to x^0$, and suppose
$\overline{\lim} \mu_n = \mu^* > \mu^0$. Then on an infinite subsequence of the n's we have $\mu_n > \alpha_j v_j$,
so that for that subsequence no new terms appear on the left side of (12)
and some may be lost. The remaining old terms with $x_j^0 > 0$ approach
$\dfrac{x_j^0}{\alpha_j} \log \dfrac{\alpha_j v_j}{\mu^*} < \dfrac{x_j^0}{\alpha_j} \log \dfrac{\alpha_j v_j}{\mu^0}$ on the subsequence. Hence equation (12)
cannot be satisfied. Now suppose $\underline{\lim} \mu_n = \mu_* < \mu^0$. Then new terms do ap-
pear on the subsequence, and all the old ones remain. But now the old
terms with $x_j^0 > 0$ approach $\dfrac{x_j^0}{\alpha_j} \log \dfrac{a_j v_j}{\mu_*} > \dfrac{x_j^0}{\alpha_j} \log \dfrac{a_j v_j}{\mu_0}$, so that again
(12) is not satisfied. It follows that $\mu^n \to \mu^0$. μ is thus continuous everywhere
in x and differentiable in x if μ is not equal to any values of $\alpha_j v_j$.

On substituting (11) into equation (6) we get

$$\varphi(x) = \sum_{\alpha_j v_j > \mu} \frac{\mu x_j}{\alpha_j} + \sum_{\alpha_j v_j \leq \mu} v_j x_j.$$ (13)

Now suppose that for two values of x^0 and x^1 we have $\mu^1 \neq \mu^0$,
and that no values of $\alpha_j v_j$ either equal μ^1 or lie strictly between μ^0
and μ^1. Denote the direction from x^0 to x^1 by γ. This direction will be
fixed when we later let x^1 approach x^0. Put $\delta x = |x^1 - x^0|$. Consider
the line segment in the $(n+1)$-dimensional (x, μ) space, joining (x^0, μ^0)
to (x^1, μ^1).

We wish to compare $\varphi(x^1)$ with $\varphi(x^0)$. We do this not[9] by following
the values of $\varphi(x)$ from x^0 to x^1, that is, requiring μ always to satisfy
equation (12), but rather by following the values of the function on the
right side of (13):

$$\psi(x, \mu) = \sum_{\alpha_j v_j > \mu} \frac{\mu x_j}{\alpha_j} + \sum_{\alpha_j v_j \leq \mu} v_j x_j$$ (14)

---

[9] The reasons for employing this method will be apparent in the harder problem
treated in the following section.

where μ is *not* required to satisfy (12) but instead moves along the line segment in (x, μ) space referred to above. We have

$$\psi(x^0, \mu^0) = \varphi(x^0), \quad \psi(x^1, \mu^1) = \varphi(x^1).$$

Suppose $\mu^1 < \mu^0$. Then, for all μ satisfying $\mu^1 \leq \mu \leq \mu^0$ (Exercise 1)

$$\psi(x, \mu) = \sum_{\alpha_j v_j \geq \mu^0} \frac{\mu x_j}{\alpha_j} + \sum_{\alpha_j v_j < \mu^0} v_j x_j. \tag{15}$$

Hence

$$\varphi(x^1) - \varphi(x^0) = \psi(x^1, \mu^1) - \psi(x^0, \mu^0)$$

$$= \sum_{\alpha_j v_j \geq \mu^0} \frac{(\mu^1 - \mu^0) x_j^1}{\alpha_j} + \sum_{\alpha_j v_j \geq \mu^0} \frac{\mu^0 (x_j^1 - x_j^0)}{\alpha_j} + \tag{16}$$

$$+ \sum_{\alpha_j v_j < \mu^0} v_j (x_j^1 - x_j^0).$$

Meanwhile, we may do the same thing with the left side of (12). Writing

$$\sigma(x, \mu) = \sum_{\alpha_j v_j > \mu} \frac{x_j}{\alpha_j} \log \frac{\alpha_j v_j}{\mu},$$

we note that if $\mu^1 < \mu^0$

$$\sigma(x, \mu) = \sum_{\alpha_j v_j \geq \mu^0} \frac{x_j}{\alpha_j} \log \frac{\alpha_j v_j}{\mu} \tag{17}$$

for all $\mu^1 \leq \mu \leq \mu^0$, so that, by the elementary law of the mean,

$$0 = \sigma(x^1, \mu^1) - \sigma(x^0, \mu^0)$$

$$= \sum_{\alpha_j v_j \geq \mu^0} \left[\frac{\gamma_j}{\alpha_j} \log \frac{\alpha_j v_j}{\mu^\theta} \cdot \delta x - \frac{x_j^\theta}{\alpha_j \mu^\theta} (\mu^1 - \mu^0) \right],$$

where $\mu^1 < \mu^\theta < \mu^0$ and x^θ is between x^0 and x^1. Hence

$$\mu^1 - \mu^0 = \mu^\theta \, \delta x \cdot \left[\frac{\displaystyle\sum_{\alpha_j v_j \geq \mu^0} \frac{\gamma_j}{\alpha_j} \log \frac{\alpha_j v_j}{\mu^\theta}}{\displaystyle\sum_{\alpha_j v_j \geq \mu^0} \frac{x_j^\theta}{\alpha_j}} \right]. \tag{18}$$

On combining (18) with (16) we get, for $\mu^1 < \mu^0$,

$$\frac{\varphi(x^1) - \varphi(x^0)}{\delta x} = \mu^\theta \cdot \left[\frac{\displaystyle\sum_{\alpha_j v_j \geq \mu^0} \frac{x_j^1}{\alpha_j}}{\displaystyle\sum_{\alpha_j v_j \geq \mu^0} \frac{x_j^\theta}{\alpha_j}} \right] \cdot \sum_{\alpha_j v_j \geq \mu^0} \frac{\gamma_j}{\alpha_j} \log \frac{\alpha_j v_j}{\mu^\theta}$$

$$+ \mu^0 \sum_{\alpha_j v_j \geq \mu^0} \frac{\gamma_j}{\alpha_j} + \sum_{\alpha_j v_j < \mu^0} \gamma_j v_j. \tag{19}$$

There is a j for which $y_j^0 > 0$. For that j, from (11), both $x_j^0 > 0$ and $\alpha_j v_j > \mu^0$. Hence

$$\sum_{\alpha_j v_j > \mu^0} \frac{x_j^0}{\alpha_j} > 0,$$

and it follows that the denominator in the square bracket in (19) is bounded from zero. Hence the ratio in square brackets is near 1 if μ^1 is near μ^0.

If $\mu^1 \geqq \mu^0$, the several $\geqq$ in (19) are replaced by $>$ and $<$ by $\leqq$. The denominator in the square bracket is again bounded from zero, and the square bracket thus close to unity if μ^1 is close to μ^0. The term dropped from the sum involving the logarithm is near zero. The sum of the last two terms is unaltered. Hence we may take the limit in (19) as $x^1 \to x^0$ and obtain

$$D_\gamma \varphi(x) = \mu^0 \sum_{\alpha_j v_j > \mu^0} \frac{\gamma_j}{\alpha_j} \left(1 + \log \frac{\alpha_j v_j}{\mu_0}\right) + \sum_{\alpha_j v_j \leqq \mu^0} \gamma_j v_j. \tag{20}$$

[*Remark.* The reader should (Exercise 2) compare formula (20) with the results gotten by attempting the lim sup and lim inf estimates by forming differences and dropping terms as in the proof of Theorem I of Chapter III. We find first

$$\limsup \frac{\varphi(x^n) - \varphi(x^0)}{d^n} \leqq \limsup \frac{H(x^n, y^0) - H(x^0, y^0)}{d^n} \leqq$$

$$\leqq \mu^0 \sum_{\substack{\alpha_j v_j > \mu^0 \\ x_j^0 > 0}} \frac{\gamma_j}{\alpha_j} \left(1 + \log \frac{\alpha_j v_j}{\mu^0}\right) + \sum_{\substack{\alpha_j v_j > \mu^0 \\ x_j = 0}} \gamma_j v_j + \sum_{\alpha_j v_j \leqq \mu^0} \gamma_j v_j. \tag{21}$$

Trying to follow the second part of that proof, handling the terms with $x_j = 0$ separately, we get

$$\liminf \frac{\varphi(x^m) - \varphi(x^0)}{d^m} \geqq \liminf \frac{H(x^m, y^m) - H(x^0, y^m)}{d^m} \geqq$$

$$\geqq \mu^0 \sum_{\substack{\alpha_j v_j > \mu^0 \\ x_j^0 > 0}} \frac{\gamma_j}{\alpha_j} \left(1 + \log \frac{\alpha_j v_j}{\mu^0}\right) + \mu^0 \sum_{\substack{\alpha_j v_j > \mu^0 \\ x_j^0 = 0}} \frac{\gamma_j}{\alpha_j} + \sum_{\alpha_j v_j \leqq \mu^0} \gamma_j v_j. \tag{22}$$

These differ from each other and from (20) when $\alpha_j v_j > \mu^0$, $x_j^0 = 0$. Thus the continuity of the partial derivatives at the origin assumed in Chapter III seems to be an important restriction. Observe that for the terms with $\alpha_j v_j > \mu^0$, $x_j^0 = 0$,

$$\frac{\mu^0}{\alpha_j} < \frac{\mu^0}{\alpha_j} \left(1 + \log \frac{\alpha_j v_j}{\mu^0}\right) < v_j, \tag{23}$$

which verifies that the estimates given by (21) and (22) do indeed enclose the correct formula (20). The second inequality of (23) follows from the fact that $\mu\left(1 + \log\dfrac{\alpha_j v_j}{\mu}\right)$ is a strictly increasing function of μ and equals v_j when $\mu = \alpha_j v_j$.]

Thus we have, by methods approaching brute force, found the formula for the directional derivative $D_\gamma \varphi(x)$. The right side of formula (20) is continuous (even when some $x_i = 0$) in x and γ; hence $D_\gamma \varphi(x)$ is continuous in those variables. The chain rule applies as in Theorem II of Chapter III, so that the directional derivative may be reduced to derivatives along the axes. For our purposes it is easier to stay with the directional derivative. We shall now study the behavior of this derivative. We shall prove that it is non-decreasing as x moves along a segment in the direction γ, and hence that φ is convex; this solves the problem since then it follows that solutions are on corners, so that only one type of system will appear in the mix.

We need a slight generalization of the lemma used to prove our law of the mean (Theorem VI in Chapter III).

Lemma 1. *If $f(x)$ is continuous on $[0, 1]$ and if the supremum of the right difference quotients at each point of $[0, 1)$ is non-negative, and if $f(0) = 0$, then $f(1) \geq 0$.*

The proof of this lemma follows closely the proof of the lemma of Chapter III, and may be left as an exercise (Exercise 3).

Let us compare the values of $D_\gamma \varphi(x)$ at two points x^0 and x^1, chosen sufficiently close that no value of $\alpha_j v_j$ equals μ^1 or lies between μ^0 and μ^1, though some $\alpha_j v_j$ may be equal to μ^0. We take x^1 to be in the γ direction from x^0. The object is to study the right difference quotients in the sense of Lemma 1. Suppose $\mu^1 < \mu^0$. Then, for (x, μ) on the segment joining (x^0, μ^0) to (x^1, μ^1) we may write the right side of formula (20) in the form

$$\Omega(x, \mu) = \mu \sum_{\alpha_j v_j \geq \mu^0} \frac{\gamma_j}{\alpha_j}\left(1 + \log\frac{\alpha_j v_j}{\mu}\right) + \sum_{\alpha_j v_j < \mu^0} \gamma_j v_j. \tag{24}$$

We thus get, using the elementary law of the mean,

$$\frac{D_\gamma \varphi(x^1) - D_\gamma \varphi(x^0)}{\delta x} = \frac{\Omega(x^1, \mu^1) - \Omega(x^0, \mu^0)}{\delta x}$$

$$= \frac{\mu^1 - \mu^0}{\delta x} \sum_{\alpha_j v_j \geq \mu^0} \frac{\gamma_j}{\alpha_j} \log\frac{\alpha_j v_j}{\mu^{\theta'}} \tag{25}$$

where $\mu^{\theta'} = (1 - \theta')\mu^0 + \theta'\mu^1$ for some $0 < \theta' < 1$. Using (18), we get

$$\frac{D_\gamma \varphi(x^1) - D_\gamma \varphi(x^0)}{\delta x} = \frac{\left[\displaystyle\sum_{\alpha_j v_j \geq \mu^0} \frac{\gamma_j}{\alpha_j} \log\frac{\alpha_j v_j}{\mu^\theta}\right]\left[\displaystyle\sum_{\alpha_j v_j \geq \mu^0} \frac{\gamma_j}{\alpha_j} \log\frac{\alpha_j v_j}{\mu^{\theta'}}\right]}{\displaystyle\sum_{\alpha_j v_j \geq \mu^0} \frac{x_j^\theta}{\alpha_j \mu^\theta}}. \tag{26}$$

If $\mu^1 \geq \mu^0$, (26) is replaced by the same formula with the expression $\alpha_j v_j \geq \mu^0$ replaced by $\alpha_j v_j > \mu^0$ throughout. As in previous calculations, the denominator on the right side of (26) is bounded from zero. The two terms in brackets approach the same number as $x^1 \to x^0$ from the direction γ. Hence the quotients on the right side of (26) have a non-negative supremum as $x^1 \to x^0$. Hence we may apply Lemma 1 to prove that $D_\gamma \varphi(x)$ is non-decreasing in the direction γ. We have thus proved the following.

Lemma 2. *For any direction γ, $D_\gamma \varphi(x)$ is non-decreasing in the direction γ.*

But this proves the more familiar statement following.

Lemma 3. $\varphi(x)$ *is convex in x.*

This follows from the elementary connection between the derivative and convexity. We give the proof here. Suppose we are given two distinct points x^0 and x^1. Let the direction from x^0 to x^1 be γ. φ is differentiable in the ordinary sense as a function of points on that segment. This follows from (20), which shows that $D_\gamma \varphi(x)$ exists everywhere and $D_{-\gamma} \varphi(x) = -D_\gamma \varphi(x)$. Write $x^\lambda = (1-\lambda) x^0 + \lambda x^1$, $0 < \lambda < 1$. Then we have the inequality

$$\varphi(x^1) - \varphi(x^\lambda) = D_\gamma \varphi(x^*) |x^1 - x^\lambda|$$
$$\geq D_\gamma \varphi(x^\lambda) |x^1 - x^\lambda| \qquad (27)$$
$$= (1-\lambda) D_\gamma \varphi(x^\lambda) |x^1 - x^0|,$$

where x^* is between x^λ and x^1. By the same reasoning we find

$$\varphi(x^\lambda) - \varphi(x^0) \leq \lambda D_\gamma \varphi(x^\lambda) |x^1 - x^0|. \qquad (28)$$

On multiplying (28) by $(1-\lambda)$, (27) by $-\lambda$, and adding, we get

$$(1-\lambda) \varphi(x^0) + \lambda \varphi(x^1) \geq \varphi(x^\lambda), \qquad (29)$$

i.e. the inequality defining convexity. The lemma is proved.

We know from the elementary theory of convex functions (essentially Exercise 6, Chapter III) that there are solutions $x_1, \ldots, x_n$ on corners. The main problem now is to discover under what conditions there exists non-extreme solutions.

Suppose given such a solution. Since $y_j = 0$ if $x_j = 0$, and $\Sigma y_j = 1$, there is at least one value of j with both $x_j > 0$ and $y_j > 0$. Suppose also $x_k > 0$. We construct a segment with the direction γ defined by $\gamma_j = -1/\sqrt{2}$, $\gamma_k = 1/\sqrt{2}$ through the point and extend it as far as possible in both directions. Denote the two end points of this segment by x^0 and x^1, and denote our solution by $x^\lambda = (1-\lambda) x^0 + \lambda x^1$, $0 < \lambda < 1$. Then $\varphi(x^\lambda) \leq (1-\lambda) \varphi(x^0) + \lambda \varphi(x^1)$. Since x^λ is a solution, $\varphi(x^\lambda) \geq \varphi(x^0)$. and $\varphi(x^\lambda) \geq \varphi(x^1)$. Hence $\varphi(x^\lambda) = \varphi(x^0) = \varphi(x^1)$. Suppose x is between x^λ and x^1, so that for some θ, $0 < \theta < 1$, $x^\lambda = (1-\theta) x^0 + \theta x$. Then $\varphi(x^\lambda) \leq$

$\leqq (1-\theta)\varphi(x^0)+\theta\varphi(x)$, from which it follows that $\varphi(x)\geqq\varphi(x^\lambda)$ and therefore $\varphi(x)=\varphi(x^\lambda)$. Using the similar argument for x between x^0 and x^λ, we find that $\varphi(x)=$ constant on the entire segment. Hence $D_y\varphi(x)=0$ there.

From (26) we see that on the entire segment we must have

$$\sum_{\alpha_{j*}v_{j*}>\mu}\frac{\gamma_{j*}}{\alpha_{j*}}\log\frac{\alpha_{j*}v_{j*}}{\mu}=0. \tag{30}$$

It follows from (18) that μ is constant along the segment. Since $y_j>0$, $\alpha_j v_j>\mu$. We must have $\alpha_k v_k>\mu$, for otherwise, from (30),

$$-\frac{1}{\sqrt{2}\alpha_j}\log\frac{\alpha_j v_j}{\mu}=0,$$

which is not so. Thus we may write (30) in the form

$$\frac{1}{\alpha_j}\log\frac{\alpha_j v_j}{\mu}=\frac{1}{\alpha_k}\log\frac{\alpha_k v_k}{\mu}. \tag{31}$$

Using (30) in (20), we get

$$\mu\sum_{\alpha_{j*}v_{j*}>\mu}\frac{\gamma_{j*}}{\alpha_{j*}}+\sum_{\alpha_{j*}v_{j*}\leqq\mu}\gamma_j v_j=0,$$

which in the case at hand reads

$$\frac{1}{\alpha_j}=\frac{1}{\alpha_k}. \tag{32}$$

We thus have $\alpha_j=\alpha_k$, and, from (31) $v_j=v_k$. Denote now by J the set of all j such that $x_j>0$. We have just proved that

$$\alpha_j=\alpha_J=\text{const.},\quad v_j=v_J=\text{const.},\qquad j\in J.$$

Applying now formula (11), we get

$$y_j=\frac{x_j}{\alpha_J}\log\frac{\alpha_J v_J}{\mu}\quad j\in J,$$
$$y_j=0\qquad\qquad j\notin J. \tag{33}$$

Now since $\Sigma y_j=\Sigma x_j=1$, (33) implies that

$$1=\frac{1}{\alpha_J}\log\frac{\alpha_J v_J}{\mu} \tag{34}$$

and hence that

$$y_j=x_j. \tag{35}$$

Conversely suppose J is a set with $\alpha_j=\alpha_J$, $v_j=v_J$ and $v_j e^{-\alpha_J}$
$=\underset{j=1,\ldots,n}{\text{Max}}\ v_j e^{-\alpha_j}=M$. Consider any set with $\Sigma_J x_j=1$, $x_j\geqq0$. We have

again (33) and hence (35), so that

$$\sum_{j=1}^{n} v_j x_j e^{-\alpha_j y_j / x_j} = v_J e^{-\alpha_J} = M.$$ (36)

Since there are corner solutions, so that the value of the Max-Min is M, (36) shows that the set x is a solution. We have thus completely proved the following theorem.

Theorem I (The second special problem). *Let J_0 be the set of those j's for which $v_j e^{-\alpha_j} = \text{Max} = M$. The corner solutions are obtained by taking, for some $j \in J_0$, $x_j = 1$ and $x_k = 0$ for all $k \neq j$. There are non-corner solutions if and only if for some subset $J \subset J_0$ we have $\alpha_j = \alpha_J = \text{constant}$, $v_j = v_J = \text{constant}$. In that case any set $x_1, \ldots, x_n$ with $\sum_{j \in J} x_j = 1$ is a solution, the answering y's being given by $y_j = x_j$.*

Thus only one really distinct weapons system will appear in the optimal mix. Theorem I thus completely solves the special problem.

We will carry through similar calculations in Section 2 to obtain the corresponding result for the mixed problem. The latter, though they are in some ways analogous and use some details from the above results, do not follow from them, since the mixed problem does not decompose into the two special problems. The general case is considerably more complicated; and this fact has justified our doing the above, simpler, case first as a guide.

Note. While going through the mixed problem, the reader is advised to carry through the entire computation with the j-terms missing. In this way he will see what the new essential difficulties are (Exercise 4).

2. The mixed problem. We turn now to the mixed problem given by (1), (3), and (4). Given first a set $x_1, \ldots, x_n, x_{n+1}, \ldots, x_m$ not all zero and satisfying $x_i, x_j \geq 0$, we find the y that yields the minimum. There exists by the Gibbs lemma a μ such that

$$v_i \alpha_i x_i e^{-\alpha_i y_i} = \mu \quad \text{if} \quad y_i > 0$$ (37)

$$v_i \alpha_i x_i e^{-\alpha_i y_i} \leq \mu \quad \text{if} \quad y_i = 0$$ (38)

$$v_j \alpha_j e^{-\alpha_j y_j / x_j} = \mu \quad \text{if} \quad x_j > 0 \quad \text{and} \quad y_j > 0$$ (39)

$$v_j \alpha_j e^{-\alpha_j y_j / x_j} \leq \mu \quad \text{if} \quad x_j > 0 \quad \text{and} \quad y_j = 0$$ (40)

$$0 = \mu \quad \text{if} \quad x_j = 0 \quad \text{and} \quad y_j > 0$$ (41)

$$0 \leq \mu \quad \text{if} \quad x_j = 0 \quad \text{and} \quad y_j = 0.$$ (42)

Since some x_i or x_j is positive, it follows from (37), (38), (39) and (40) that $\mu > 0$. From (41) it then follows that $x_j = 0$ implies $y_j = 0$, and from

(37) that $x_i = 0$ implies $y_i = 0$. We may then characterize the solution as follows:

$$y_i = \frac{1}{\alpha_i} \log \frac{\alpha_i v_i x_i}{\mu} \quad \text{if} \quad \alpha_i v_i x_i > \mu,$$
$$= 0 \quad \text{otherwise};$$

(43)

$$y_j = \frac{x_j}{\alpha_j} \log \frac{\alpha_j v_j}{\mu} \quad \text{if} \quad \alpha_j v_j > \mu,$$
$$= 0 \quad \text{otherwise}.$$

(44)

This gives us the formula, analogous to (20),

$$\varphi(x) = \sum_{\alpha_i v_i x_i > \mu} \frac{\mu}{\alpha_i} + \sum_{\alpha_i v_i x_i \leq \mu} v_i x_i + \sum_{\alpha_j v_j > \mu} \frac{\mu x_j}{\alpha_j} + \sum_{\alpha_j v_j \leq \mu} v_j x_j,$$

(45)

the μ being the solution of

$$\sum_{\alpha_i v_i x_i > \mu} \frac{1}{\alpha_i} \log \frac{\alpha_i v_i x_i}{\mu} + \sum_{\alpha_j v_j > \mu} \frac{x_j}{\alpha_j} \log \frac{\alpha_j v_j}{\mu} = 1.$$

(46)

Since the left hand side of (46) is continuous in μ and decreases from large values to zero as μ increases, then (46) has a unique solution. By an argument analogous to that following formula (12), μ is a continuous function of x.

Denote the right side of (45), for μ free, by $\psi(x, \mu)$. Denote the left side of (46) by $\sigma(x, \mu)$. Take x^1 close to x^0. Let μ^0 and μ^1 correspond to x^0 and x^1 respectively. Choose x^1 so that:

(i) For each i, if $\alpha_i v_i x_i^0 \neq \mu^0$ then $\alpha_i v_i x_i - \mu$ remains of the same sign throughout the closed interval joining (x^0, μ^0) to (x^1, μ^1).

(ii) For each j, if $\alpha_j v_j \neq \mu^0$ then $\alpha_j v_j - \mu$ remains of the same sign throughout that interval.

We wish again to calculate $\varphi(x^1) - \varphi(x^0)$. Denote the direction from x^0 to x^1 by γ. Put $|x^1 - x^0| = \delta x$. Suppose $\mu^1 < \mu^0$. Use Σ^* to denote the sum over the set of i's for which either $\alpha_i v_i x_i^0 > \mu^0$ or $\alpha_i v_i x_i^0 = \mu^0$ and $\alpha_i v_i \gamma_i > \dfrac{\mu^1 - \mu^0}{\delta x}$. Let Σ^{**} denote the complementary set of i's. Then for all (x, μ) on the segment joining (x^0, μ^0) to (x^1, μ^1) we have

$$\psi(x, \mu) = \Sigma^* \frac{\mu}{\alpha_i} + \Sigma^{**} v_i x_i +$$
$$+ \sum_{\alpha_j v_j \geq \mu^0} \frac{\mu x_j}{\alpha_j} + \sum_{\alpha_j v_j < \mu^0} v_j x_j.$$

(47)

Thus we get

$$\varphi(x^1) - \varphi(x^0) = \psi(x^1, \mu^1) - \psi(x^0, \mu^0)$$

$$= \Sigma^* \frac{\mu^1 - \mu^0}{\alpha_i} + \Sigma^{**} v_i(x_i^1 - x_i^0) +$$

$$+ \sum_{\alpha_j v_j \geq \mu^0} \frac{\mu^1 - \mu^0}{\alpha_j} x_j^1 + \sum_{\alpha_j v_j \geq \mu^0} \frac{\mu^0}{\alpha_j}(x_j^1 - x_j^0) + \tag{48}$$

$$+ \sum_{\alpha_j v_j < \mu^0} v_j(x_j^1 - x_j^0).$$

On rewriting (48) we get

$$\frac{\varphi(x^1) - \varphi(x^0)}{\delta x} = \frac{\mu^1 - \mu^0}{\delta x}\left[\Sigma^* \frac{1}{\alpha_i} + \sum_{\alpha_j v_j \geq \mu^0} \frac{x_j^1}{\alpha_j}\right] +$$

$$+ \Sigma^{**}\gamma_i v_i + \mu^0 \sum_{\alpha_j v_j \geq \mu^0} \frac{\gamma_j}{\alpha_j} + \sum_{\alpha_j v_j < \mu^0} \gamma_j v_j. \tag{49}$$

To get the relation between $\mu^1 - \mu^0$ and δx, we observe that for (x, μ) on the segment joining (x^0, μ^0) to (x^1, μ^1)

$$\sigma(x, \mu) = \Sigma^* \frac{1}{\alpha_i} \log \frac{\alpha_i v_i x_i}{\mu} + \sum_{\alpha_j v_j \geq \mu^0} \frac{x_j}{\alpha_j} \log \frac{\alpha_j v_j}{\mu}. \tag{50}$$

The star has the same meaning as in formula (47). We thus calculate, the θ indicating an intermediate point as before,

$$0 = \sigma(x^1, \mu^1) - \sigma(x^0, \mu^0) = \delta x \left[\Sigma^* \frac{\gamma_i}{\alpha_i x_i^\theta} + \sum_{\alpha_j v_j \geq \mu^0} \frac{\gamma_j}{\alpha_j} \log \frac{\alpha_j v_j}{\mu^\theta}\right] -$$

$$- (\mu^1 - \mu^0)\left[\Sigma^* \frac{1}{\alpha_i \mu^\theta} + \sum_{\alpha_j v_j \geq \mu^0} \frac{x_j^\theta}{\alpha_j \mu^\theta}\right]. \tag{51}$$

On combining (49) with (51) we get, for $\mu^1 < \mu^0$,

$$\frac{\varphi(x^1) - \varphi(x^0)}{\delta x} = \mu^\theta \times \frac{\left[\Sigma^* \dfrac{1}{\alpha_i} + \sum\limits_{\alpha_j v_j \geq \mu^0} \dfrac{x_j^1}{\alpha_j}\right]}{\left[\Sigma^* \dfrac{1}{\alpha_i} + \sum\limits_{\alpha_j v_j \geq \mu^0} \dfrac{x_j^\theta}{\alpha_j}\right]} \times$$

$$\times \left[\Sigma^* \frac{\gamma_i}{\alpha_i x_i^\theta} + \sum_{\alpha_j v_j \geq \mu^0} \frac{\gamma_j}{\alpha_j} \log \frac{\alpha_j v_j}{\mu^\theta}\right] + \tag{52}$$

$$+ \Sigma^{**}\gamma_i v_i + \mu^0 \sum_{\alpha_j v_j \geq \mu^0} \frac{\gamma_j}{\alpha_j} + \sum_{\alpha_j v_j < \mu^0} \gamma_j v_j.$$

If $\mu^1 \geq \mu^0$ formula (52) is replaced by one in which $\alpha_j v_j \geq \mu^0$ is replaced by $\alpha_j v_j > \mu^0$ and $\alpha_j v_j < \mu^0$ by $\alpha_j v_j \leq \mu^0$. As $x^1 \to x^0$ from the direction γ, both formulas have the same limit (Exercise 5). Hence we obtain finally the general formula for the directional derivative in the direction γ:

$$
D_\gamma \varphi(x) = \mu \sum_{\alpha_i v_i x_i > \mu} \frac{\gamma_i}{\alpha_i x_i} + \sum_{\alpha_i v_i x_i \leq \mu} \gamma_i v_i +
$$
$$
+ \mu \sum_{\alpha_j v_j > \mu} \frac{\gamma_j}{\alpha_j} \left(1 + \log \frac{\alpha_j v_j}{\mu} \right) + \sum_{\alpha_j v_j \leq \mu} \gamma_j v_j. \tag{53}
$$

We now wish to compare the values of $D_\gamma \varphi(x)$ at two neighboring points x^0 and x^1, with x^1 in the direction γ from x^0. Suppose $\mu^1 < \mu^0$. We again construct a segment from (x^0, μ^0) to (x^1, μ^1) and write the right side of (53), for (x, μ) on that segment [μ not constrained by (46), except at the end points] in the form

$$
\Omega(x, \mu) = \mu \Sigma^* \frac{\gamma_i}{\alpha_i x_i} + \Sigma^{**} \gamma_i v_i +
$$
$$
+ \mu \sum_{\alpha_j v_j \geq \mu^0} \frac{\gamma_j}{\alpha_j} \left(1 + \log \frac{\alpha_j v_j}{\mu} \right) + \sum_{\alpha_j v_j < \mu^0} \gamma_j v_j, \tag{54}
$$

the stars having the same meaning as in formula (47). We have, by the elementary law of the mean, for some θ' with $0 < \theta' < 1$,

$$
D_\gamma \varphi(x^1) - D_\gamma \varphi(x^0) = \Omega(x^1, \mu^1) - \Omega(x^0, \mu^0)
$$
$$
= (\mu^1 - \mu^0) \left[\Sigma^* \frac{\gamma_i}{\alpha_i x_i^{\theta'}} + \sum_{\alpha_j v_j \geq \mu^0} \frac{\gamma_j}{\alpha_j} \log \frac{\alpha_j v_j}{\mu^{\theta'}} \right] -
$$
$$
- \mu^{\theta'} \Sigma^* \frac{\gamma_i \, \delta x}{\alpha_i (x_i^{\theta'})^2} (x_i^1 - x_i^0). \tag{55}
$$

Substituting $\mu^1 - \mu^0$ from (51) into (55), we get, for $\mu^1 < \mu^0$,

$$
\frac{D_\gamma \varphi(x^1) - D_\gamma \varphi(x^0)}{\delta x}
$$
$$
= \frac{\left[\Sigma^* \frac{\gamma_i}{\alpha_i x_i^{\theta'}} + \sum_{\alpha_j v_j \geq \mu^0} \frac{\gamma_j}{\alpha_j} \log \frac{\alpha_j v_j}{\mu^{\theta'}} \right] \left[\Sigma^* \frac{\gamma_i}{\alpha_i x_i^{\theta}} + \sum_{\alpha_j v_j \geq \mu^0} \frac{\gamma_j}{\alpha_j} \log \frac{\alpha_j v_j}{\mu^{\theta}} \right]}{\Sigma^* \frac{1}{\alpha_i \mu^{\theta}} + \sum_{\alpha_j v_j \geq \mu^0} \frac{x_j^{\theta}}{\alpha_j \mu^{\theta}}} -
$$
$$
- \mu^{\theta'} \Sigma^* \frac{\gamma_i^2}{\alpha_i (x_i^{\theta'})^2}. \tag{56}
$$

A similar formula holds for $\mu^1 \geq \mu^0$. We may now state a principal result.

Theorem II (The general problem). *For directions γ with the $\gamma_j = 0$, $\varphi(x)$ is concave (not necessarily strictly) in the set of variables x_i. For directions γ with $\gamma_i = 0$, $\varphi(x)$ is convex (not necessarily strictly) in the of variables x_j.*

Proof. We prove the first part. Then the sums involving γ_j are zero.

The two sums in brackets involving x_i^θ and $x_i^{\theta'}$ have the following property. As $x^1 \to x^0$ from the direction γ the number of terms in each sum may vary, but the individual terms approach each other, and the terms present at any stage are the same. Two cases can happen.

(i) *The left sum approaches zero.* Then so does the right sum, so that

$$\limsup_{x^1 \to x^0} \frac{D_\gamma \varphi(x^1) - D_\gamma \varphi(x^0)}{\delta x} \leqq 0. \tag{57}$$

(ii) *The left sum is bounded from zero on a sequence of values x^p of x^1 as $x^p \to x^0$.* Then the ratio R_p of the right sum by the left sum approaches unity, and we may write (noting that $x_j^\theta = x_j^0$)

$$\frac{D_\gamma \varphi(x^p) - D_\gamma \varphi(x^0)}{\delta x} \tag{58}$$

$$= \frac{R^p \left(\Sigma^* \dfrac{\gamma_i}{\alpha_i x_i^{\theta'}}\right)^2 - \dfrac{\mu^{\theta'}}{\mu^\theta} \Sigma^* \dfrac{1}{\alpha_i} \Sigma^* \dfrac{\gamma_i^2}{\alpha_i (x_i^{\theta'})^2} - \dfrac{\mu^{\theta'}}{\mu^\theta} \displaystyle\sum_{\alpha_j v_j \geqq \mu^0} \dfrac{x_j^0}{\alpha_j} \Sigma^* \dfrac{\gamma_i^2}{\alpha_i (x_i^{\theta'})^2}}{\Sigma^* \dfrac{1}{\alpha_i \mu^\theta} + \displaystyle\sum_{\alpha_j v_j \geqq \mu^0} \dfrac{x_j^0}{\alpha_j \mu^\theta}}.$$

Using the Schwartz inequality (Exercise 6), we obtain

$$\left(\Sigma^* \frac{\gamma_i}{\alpha_i x_i^{\theta'}}\right)^2 = \left(\Sigma^* \frac{\gamma_i}{\alpha_i^{1/2} x_i^{\theta'}} \cdot \frac{1}{\alpha_i^{1/2}}\right)^2 \leqq \Sigma^* \frac{\gamma_i^2}{\alpha_i (x_i^{\theta'})^2} \cdot \Sigma^* \frac{1}{\alpha_i}.$$

Applying this to (58) and dropping the last term in the numerator, we get

$$\frac{D_\gamma \varphi(x^p) - D_\gamma \varphi(x^0)}{\delta x} \leqq \frac{\left(R^p - \dfrac{\mu^{\theta'}}{\mu^\theta}\right)\left(\Sigma^* \dfrac{\gamma_i^2}{\alpha_i x_i^{\theta'}}\right)}{\Sigma^* \dfrac{1}{\alpha_i \mu^\theta} + \displaystyle\sum_{\alpha_j v_j \geqq \mu^0} \dfrac{x_j^0}{\alpha_j \mu^\theta}}.$$

But $\mu^\theta/\mu^{\theta'} \to 1$ and $R_p \to 1$, and the coefficient of R_p is bounded, so that

$$\limsup_{x^p \to x^0} \frac{D_\gamma \varphi(x^p) - D_\gamma \varphi(x^0)}{\delta x} \leqq 0.$$

Now consider the γ direction to be that of the positive real axis. We have proved that the supremum and hence *a fortiori* the infimum of the right difference quotients of $D_\gamma \varphi(x)$ is non-positive. Applying Lemma 1, we find that $D_\gamma \varphi(x)$ is non-increasing in the direction γ. Hence $\varphi(x)$ is concave in directions for which $\gamma_j = 0$.

The proof of the second statement, is as in the proof of Lemma 3, immediate from Lemma 1.

Theorem III (The general problem). *Let x be a solution to the general problem and let y be the corresponding solution to the inside problem. Put*

$$\sum_{j=n+1}^{m} x_j = \xi; \quad \sum_{j=n+1}^{m} y_j = \eta. \tag{59}$$

Let J be the set of j's for which $x_j > 0$. Then if $j \in J$ $v_j = \text{constant} = v_J$. If in addition $\eta > 0$ we have $\alpha_j = \text{constant} = \alpha_J$ as well, for $j \in J$. Conversely suppose J is a set all of whose elements yield

$$\underset{n+1 \leq j \leq m}{\text{Max}} \, v_j e^{-\alpha_j \eta/\xi} = M$$

for the values of ξ and η given by (59). Suppose further that $v_j = v_J$ for $j \in J$, and that $\alpha_j = \alpha_J$ for $j \in J$ if $\eta > 0$. Then any vector x^* with $x_i^* = x_i$ for all i and with $\sum_{j \in J} x_j^* = \xi$ also yields a solution, with the corresponding

y^* satisfying $y_i^* = y_i$ for all i and $y_j^* = \dfrac{\eta x_j^*}{\xi}$ for $j \in J$.

The proof, which is an easy variation on the proof of Theorem I, is left as an exercise for the reader (Exercise 7).

3. The balance between systems of types i and j

This section treats the fundamental question: under what conditions are the solutions "of type i," i.e. all $x_j = 0$, or "of type j," i.e. all $x_i = 0$, or "mixed," i.e. some $x_i > 0$ and some $x_j > 0$?

This problem is treated as follows: first, as follows from Theorem III, it is only necessary to consider one j at a time. This understood, we solve the problem with an additional artifical constraint, in which a proportion ξ of the x-player's resources is arbitrarily assigned to the jth system. The resulting value of the Max-Min is then $H(\xi)$. The shape of this function is studied in detail. This is a complicated proceeding, and culminates in a "Description of the Behavior of $H(\xi)$" beginning just above formula (103). From this Description one may calculate the ξ yielding the maximum.

From this point on we shall fix on some j. We shall write for convenience $\xi = x_j$ and $\eta = y_j$. The letter x will from now on represent the set of x_i's only, and similarly for y. Now put

$$\varphi_j(x, \xi) = \underset{y, \eta}{\text{Min}} \left[\Sigma v_i x_i e^{-\alpha_i y_i} + v_j \xi e^{-\alpha_j \eta/\xi} \right], \tag{60}$$

the minimum being taken subject to the constraint

$$\Sigma y_i + \eta = 1, \quad y_i \geq 0, \quad \eta \geq 0. \tag{61}$$

We shall from now on regard ξ as an independent parameter. For ξ fixed, define

$$H(\xi) = \underset{x}{\text{Max}} \, \varphi_j(x, \xi), \tag{62}$$

where

$$\Sigma x_i = 1 - \xi, \quad x_i \geq 0. \tag{63}$$

In the rest of this chapter we shall be studying the properties of $H(\xi)$. Suppose ξ^0 yields the maximum to $H(\xi)$ in the interval $[0, 1]$, and that x^0 corresponds. Then the set $(x^0, \xi) = (x_1^0, \ldots, x_n^0, \xi)$ is the solution for the particular value of j in question. On running through the j's one thus solves the original problem (1)—(4).

Now $\varphi_j(x, \xi)$ is continuously differentiable in x and in ξ. In fact formula (53) gives

$$D_\gamma \varphi_j(x, \xi) = \sum_{\alpha_i v_i x_i > \mu} \frac{\mu \gamma_i}{\alpha_i x_i} + \sum_{\alpha_i v_i x_i \leq \mu} \gamma_i v_i, \tag{64}$$

for directions γ in the x-space (i.e. in the current notation with $\gamma_j = 0$), and this is evidently continuous [as is so in the general case for (53) as well]. Thus the elementary calculus applies and we may use the trivial variant of the Gibbs Lemma to find that if x yields the maximum in (62) subject to (63), then for some λ depending on ξ we have

$$\frac{\mu}{\alpha_i x_i} = \lambda \quad \text{if} \quad \alpha_i v_i x_i > \mu; \tag{65}$$

$$v_i = \lambda \quad \text{if} \quad 0 < \alpha_i v_i x_i \leq \mu; \tag{66}$$

$$v_i \leq \lambda \quad \text{if} \quad x_i = 0. \tag{67}$$

Here the partial derivatives of $\varphi_j(x, \xi)$ with respect to the x_i were obtained by inspection from (64). If $\xi = 1$, λ is not determined. We need to recall the source of the μ in formula (53), which led to formula (64). This was given by formula (46), which in turn resulted from the formulas for y and η in terms of x and ξ given by (43) and (44). We recall these:

$$y_i = \frac{1}{\alpha_i} \log \frac{\alpha_i v_i x_i}{\mu} \quad \text{if} \quad \alpha_i v_i x_i > \mu;$$
$$= 0 \quad \text{otherwise}. \tag{68}$$

$$\eta = \frac{\xi}{\alpha_j} \log \frac{\alpha_j v_j}{\mu} \quad \text{if} \quad \alpha_j v_j > \mu;$$
$$= 0 \quad \text{otherwise}. \tag{69}$$

We leave as an exercise (Exercise 8) for the reader to deduce from (65), (66), (67), (68) and (69) the following characterization of the solution:

$$v_i > \lambda: \quad \begin{aligned} x_i &= \frac{\mu}{\alpha_i \lambda}, \tag{70} \\ y_i &= \frac{1}{\alpha_i} \log \frac{v_i}{\lambda}; \tag{71} \end{aligned}$$

$$v_i = \lambda: \quad x_i \leq \frac{\mu}{\alpha_i \lambda}, \quad y_i = 0; \tag{72}$$

$$v_i < \lambda: \quad x_i = y_i = 0;$$

$$\alpha_j v_j > \mu: \quad \eta = \frac{\xi}{\alpha_j} \log \frac{\alpha_j v_j}{\mu}; \tag{73}$$

$$\alpha_j v_j \leq \mu: \quad \eta = 0.$$

The reader should also observe (Exercise 9) that the problem (60), (62) is, for each fixed ξ, a game. Then (70)—(73) will follow from the game conditions.

From (70) and (63), we get

$$\sum_{v_i > \lambda} \frac{\mu}{\lambda \alpha_i} + \sum_{v_i = \lambda} x_i = 1 - \xi. \tag{74}$$

From (71) and the side condition $\Sigma y_i + \eta = 1$, we find

$$\sum_{v_i > \lambda} \frac{1}{\alpha_i} \log \frac{v_i}{\lambda} = 1 - \eta. \tag{75}$$

Writing

$$\omega(\lambda) = \sum_{v_i > \lambda} \frac{1}{\alpha_i}, \quad \zeta(\lambda) = \sum_{v_i > \lambda} \frac{1}{\alpha_i} \log \frac{v_i}{\lambda}, \tag{76}$$

we may rewrite (74) and (75) in the form

$$\frac{\mu \omega(\lambda)}{\lambda} + \sum_{v_i = \lambda} x_i = 1 - \xi; \tag{77}$$

$$\zeta(\lambda) = 1 - \eta. \tag{78}$$

Finally, we list (73) again:

$$\eta = \frac{\xi}{\alpha_j} \log \frac{\alpha_j v_j}{\mu} \quad \text{if} \quad \alpha_j v_j > \mu; \quad \text{otherwise} \quad \eta = 0. \tag{79}$$

Observe that $\zeta(\lambda)$ is continuous for all λ and differentiable in λ if $\lambda \neq v_i$ for all i, and that for such values

$$\zeta'(\lambda) = -\frac{\omega(\lambda)}{\lambda}. \tag{80}$$

$\omega(\lambda)$ is of course not continuous; it is non-increasing in λ.

We take as the *a priori* range of variation of the unknowns λ, η, and μ in equations (77)—(79) the following: $\lambda > 0$, $0 \leq \eta \leq 1$, and $\mu > 0$. Recall that λ is not determined if $\xi = 1$. We shall now prove a number of lemmas about the solution of the system (77), (78) and (79) for λ, μ, and η in terms of ξ.

Lemma 4. *For each $\xi < 1$, λ is unique.*

Proof. Suppose that for two solutions of the system we have $\lambda_1 > \lambda_2$. Then, from (78) we have $\eta_1 > \eta_2$, so $\eta_1 > 0$. Hence $\alpha_j v_j > \mu_1$. Suppose that $\mu_2 \leq \mu_1$. Then $\alpha_j v_j > \mu_2$ and we have

$$\eta_2 = \frac{\xi}{\alpha_j} \log \frac{\alpha_j v_j}{\mu_2} \geq \frac{\xi}{\alpha_j} \log \frac{\alpha_j v_j}{\mu_1} = \eta_1,$$

a contradiction. Hence $\mu_2 > \mu_1$. We have therefore, using (72),

$$1 - \xi = \frac{\mu_1}{\lambda_1} \omega(\lambda_1) + \sum_{v_i = \lambda_1} x_i \leq \frac{\mu_1}{\lambda_1} \sum_{v_i \geq \lambda_1} \frac{1}{\alpha_i} < \frac{\mu_2}{\lambda_2} \sum_{v_i > \lambda_2} \frac{1}{\alpha_i} \leq 1 - \xi$$

so that $1 - \xi < 1 - \xi$, a contradiction. Hence λ is unique.

Lemma 5. η *is unique for* $\xi < 1$.

Proof. This follows from (78) and Lemma 4.

Lemma 6. *If* $1 > \xi_2 > \xi_1$, *then* $\eta_2 \geqq \eta_1$.

Proof. Suppose $\eta_2 < \eta_1$. Then $\eta_1 > 0$, so that $\alpha_j v_j > \mu_1$. Suppose $\mu_2 \leqq \mu_1$. Then $\alpha_j v_j > \mu_2$ and

$$\eta_2 = \frac{\xi_2}{\alpha_j} \log \frac{\alpha_j v_j}{\mu_2} > \frac{\xi_1}{\alpha_j} \log \frac{\alpha_j v_j}{\mu_1} = \eta_1,$$

a contradiction. Hence $\mu_2 > \mu_1$. Now from (78) we have $\lambda_2 < \lambda_1$, so that, using (72),

$$1 - \xi_1 = \frac{\mu_1}{\lambda_1} \omega(\lambda_1) + \sum_{v_i = \lambda_1} x_i \leqq \frac{\mu_1}{\lambda_1} \sum_{v_i \geqq \lambda_1} \frac{1}{\alpha_i} < \frac{\mu_2}{\lambda_2} \omega(\lambda_2) \leqq 1 - \xi_2,$$

from which $\xi_2 < \xi_1$, a contradiction. The lemma is proved.

Lemma 7. λ *and* η *are continuous functions of* ξ, *for* $\xi < 1$.

Proof. Suppose $\xi_k \to \xi^* < 1$. Suppose that x^k yields the maximum in $\varphi_j(x, \xi^k)$ and that the pair y^k, η^k minimizes against x^k, ξ^k. Then the system (70)—(73) holds for x^k, ξ^k, y^k, η^k for some λ^k and μ^k, the λ^k being, from Lemma 4, unique. Take a subsequence $\{k'\}$ such that x^k, y^k, λ^k and μ^k converge to limits x', y', λ' and μ' respectively. Then η converges to a limit η' as well, and on passing to the limit in (70—73) one easily sees that (70)—(73) hold for x', ξ^*, y', η', λ' and μ'. Hence (77)—(79) hold for ξ^*, η', λ' and μ'. Lemma 4 now implies that $\lambda' = \lambda^*$ and the proof is complete.

Lemma 8. *There is a maximal interval* $[0, \xi^*]$ *with* $\xi^* < 1$ *such that* $\eta = 0$ *on* $[0, \xi^*]$ *and* $\eta > 0$ *on* $(\xi^*, 1)$. *The value of* ξ^* *is given by*

$$\xi^* = 1 - \frac{\alpha_j v_j \omega(\lambda^0)}{\lambda^0}, \tag{81}$$

if the quantity on the right is positive and by zero otherwise, where λ^0 *corresponds to* $\xi = 0$, *i.e. is the value of the i-game alone, as in Chapter II with* $\Sigma x_i = \Sigma y_i = 1$. *In any case* $\eta \to 0$ *as* $\xi \to 0$.

Proof. Suppose first that $\eta = 0$ throughout $[0, 1)$. Then, from (78), $\zeta(\lambda)$ is a fixed positive number, so that λ is constant and some $v_i > \lambda$. It follows from (77) that μ approaches zero as ξ approaches unity. Thus, from (79), $\eta > 0$ for ξ close to 1, a contradiction. The first statement of the lemma now follows with the aid of Lemma 6.

Now suppose $\xi^* > 0$. Let $\xi^k \to \xi^*$ from the right. Then, from Lemma 3 and the definition of ξ^*, $\eta^k > 0$ for each k. It follows from (78) that $\lambda^k > \lambda^*$ for each k. Since $\lambda^k \to \lambda^*$ by Lemma 7, and since $\lambda^k > \lambda^*$, then if k is large enough λ^k is not equal to any value of v_i. Hence from (77)

$$\frac{\mu^k \omega(\lambda^k)}{\lambda^k} = 1 - \xi^k. \tag{82}$$

From (79)

$$\mu^k = \alpha_j v_j e^{-\alpha_j \eta^k / \xi^k}, \tag{83}$$

so that $\mu^k \to \alpha_j v_j$ as $k \to \infty$. Passing to the limit in (82), and taking account of the fact that $\lambda^* = \lambda^0$, we get (81).

If now $\xi^* = 0$, we again have (82). From (83) we however now deduce only that $\mu^k < \alpha_j v_j$ for each k. Thus for each k

$$\frac{\alpha_j v_j \omega(\lambda^k)}{\lambda^k} > 1 - \xi^k. \tag{84}$$

On taking the limit in (84) we get

$$\frac{\alpha_j v_i \omega(\lambda^0)}{\lambda^0} \geqq 1. \tag{85}$$

The last sentence of the lemma is trivial when $\xi^* > 0$. If $\xi^* = 0$ we note that $\omega(\lambda^k) = \omega(\lambda^0)$, for large k, so that

$$\mu^k \to \frac{\lambda^0}{\omega(\lambda^0)} \tag{86}$$

from (82). μ^k is thus bounded from zero and the result follows from (79).

Lemma 9. *There is always a non-degenerate terminal interval $[\xi^{**}, 1)$ on which $\eta = 1$ and $\lambda = \bar{v}$, where $\bar{v} = \underset{i}{\text{Max}} v_i$. The value of ξ^{**} is the unique solution of the equations*

$$\mu = \alpha_j v_j \exp\left[-\alpha_j / \xi\right] \tag{87}$$

and

$$\frac{\mu}{\bar{v}} \sum_{v_i = \bar{v}} \frac{1}{\alpha_i} = 1 - \xi \tag{88}$$

gotten by eliminating μ.

Proof. It is obvious that λ never exceeds $\bar{v}$ for values of ξ less than unity. For instance, if it did the left side of (77) would be zero and thus not equal to $1 - \xi$. Now suppose $\lambda < \bar{v}$ for values of ξ arbitrarily close to 1. Then

$$\omega(\lambda) > \sum_{v_i = \bar{v}} \frac{1}{\alpha_i}.$$

Secondly, η is positive and bounded above as $\xi \to 1$, so that μ as given by (79) is bounded below as $\xi \to 1$: $\mu \geqq \bar{\mu} > 0$. Hence

$$\frac{\mu}{\lambda} \omega(\lambda) > \frac{\bar{\mu}}{\bar{v}} \sum_{v_i = \bar{v}} \frac{1}{\alpha_i},$$

so that the left side of (77) is bounded from zero as $\xi \to 1$. It could therefore not be equal to the right side for ξ close to 1. This is a contradiction. From Lemma 6 and formula (78), λ is nondecreasing. Hence $\lambda = \bar{v}$ on a nondegenerate terminal interval. Since $\zeta(\bar{v}) = 0$ [formula (76)] it follows from (78) that $\eta = 1$ there.

As to the second statement, it is clear that (87) is satisfied at the left end ξ^{**} of the terminal interval. For points just to the left of ξ^{**} we have

$$\frac{\mu}{\lambda} \sum_{v_i > \lambda} \frac{1}{\alpha_i} = \frac{\mu}{\lambda} \sum_{v_i = \bar{v}} \frac{1}{\alpha_i} = 1 - \xi.$$

On taking the limit from the left we get (88). Since the μ given by (87) increases from zero to a positive number as ξ goes from 0 to 1, and as the μ given by (88) decreases from a positive number to zero on the same interval, (87) and (88) have a unique solution $0 < \xi^{**} < 1$.

The lemma is proved.

Lemma 10. *If* $\lambda = $ constant *on a non-degenerate subinterval of* $[\xi^*, 1)$ *then the constant is a value of* v_i. *Conversely, for any* $v_i > \lambda^0$, *there is a non-degenerate interval on which* $\lambda = v_i$.

Proof. As to the first part, η must then be constant on the same interval, from (78). By the definition of ξ^*, η is positive inside the interval, so that $\eta = \dfrac{\xi}{\alpha_j} \log \dfrac{\alpha_j v_j}{\mu}$ from (79). But then μ must be increasing inside the interval. Thus the first term in (77) is increasing. Since the right side is decreasing, the second term must be decreasing and hence not zero. Hence $\lambda = v_i$ for some i.

As to the second part, if $v_i = \bar{v} = \underset{i}{\text{Max}} v_i$, the statement follows from Lemma 9. Suppose $\lambda^0 < v_i < \bar{v}$. Then it follows from Lemma 6 that the set where $\lambda = v_i$ is an interval $[a, b]$, where $\xi^* < a \leq b < 1$. Since η is unique, continuous and positive on $(\xi^*, 1)$, we note that μ is, from (79), also unique, continuous and positive there. Now for ξ just to the left of a, $\lambda < v_i$, so that from (77),

$$\begin{aligned} 1 - \xi &= \frac{\mu}{\lambda} \omega(\lambda) = \frac{\mu}{\lambda} \sum_{v_{i'} > \lambda} \frac{1}{\alpha_{i'}} = \frac{\mu}{\lambda} \sum_{v_{i'} \geq v_i} \frac{1}{\alpha_{i'}} \\ &= \frac{\mu}{\lambda} \omega(v_i) + \frac{\mu}{\lambda} \sum_{v_{i'} = v_i} \frac{1}{\alpha_{i'}}. \end{aligned} \tag{89}$$

On taking the limit as $\xi \to a-$ in (89) we get

$$\frac{\mu(a)}{v_i} \omega(v_i) + \frac{\mu(a)}{v_i} \sum_{v_{i'} = v_i} \frac{1}{\alpha_{i'}} = 1 - a. \tag{90}$$

If ξ is just to the right of b, $\lambda > v_i$, so that, from (77),

$$1 - \xi = \frac{\mu}{\lambda} \omega(\lambda) = \frac{\mu}{\lambda} \sum_{v_{i'} > \lambda} \frac{1}{\alpha_{i'}} = \frac{\mu}{\lambda} \sum_{v_{i'} > v_i} \frac{1}{\alpha_{i'}} = \frac{\mu}{\lambda} \omega(v_i). \tag{91}$$

On taking the limit from the right in (91), we get

$$\frac{\mu(b)}{v_i}\,\omega(v_i) = 1 - b. \qquad (92)$$

(90) and (92) taken together now imply that $b > a$, and the lemma is proved.

The graphs of λ and η.

From the foregoing we see that the graphs of λ and η may be described as follows. They are continuous and non-decreasing. There may be an initial interval $[0, \xi^*]$ where $\eta = 0$ and $\lambda = \lambda_0$. On this initial interval μ is not necessarily uniquely defined; however, this is of no importance as we shall see. There is always a non-degenerate terminal interval $[\xi^{**}, 1)$ where $\lambda = \bar{v} = \underset{i}{\text{Max}}\, v_i$. When λ crosses a value v_i with $\lambda^0 < v_i < \bar{v}$, there is a non-degenerate interval $[a, b]$ on which $\lambda = v_i$. We have

$$\frac{\mu(a)}{v_i}\sum_{v_{i'} \geq v_i}\frac{1}{\alpha_{i'}} = 1 - a$$

and

$$\frac{\mu(b)}{v_i}\sum_{v_{i'} > v_i}\frac{1}{\alpha_{i'}} = 1 - b.$$

On such an interval μ is strictly increasing, since $\eta = \dfrac{\xi}{\alpha_j}\log\dfrac{\alpha_j v_j}{\mu}$.

On non-initial intervals where λ is between values of v_i both λ and η are strictly increasing and the second term on the left in (77) is missing. The fact that $\omega(\lambda)$ is constant on such an interval makes it easy to calculate the formulas for the derivatives of μ, η, and λ with respect to ξ from (77), (78), and the equation in (79). We get (Exercise 10) for $\lambda \neq v_i$,

$$\mu' = \frac{\dfrac{\mu\eta}{\xi} - \lambda}{D} \qquad (93)$$

$$\eta' = \frac{\dfrac{\eta\omega}{\xi} + \dfrac{\lambda\xi}{\alpha_j\mu}}{D} \qquad (94)$$

$$\lambda' = \frac{\lambda}{\omega(\lambda)}\,\eta' \qquad (95)$$

where

$$D = \omega + \frac{\xi}{\alpha_j}. \qquad (96)$$

Study of $H(\xi)$.

Using formulas (71) and (72) and the fact that $x_i = 0$ if $v_i < \lambda$, we compute

$$\Sigma v_i x_i e^{-\alpha_i y_i} = \sum_{v_i > \lambda} \lambda x_i + \sum_{v_i = \lambda} v_i x_i = \lambda \Sigma x_i = \lambda(1 - \xi).$$

Hence we have the basic formula

$$H(\xi) = \lambda(1 - \xi) + v_j \xi e^{-\alpha_j \eta/\xi}. \tag{97}$$

On an initial interval where $\eta = 0$ we have that $H(\xi)$ is linear and between λ^0 and v_j. (Notice that the possibly indeterminate μ on that interval does not enter.)

We now need to study $H(\xi)$ on the flat (i.e. $\lambda = v_i$) intervals and the intervals with $\lambda \neq v_j$. As to the flat intervals, the expression $\xi e^{-\alpha_j \eta/\xi}$ is strictly convex in ξ; λ and η are constants; therefore $H(\xi)$ is strictly convex on such an interval. We note that on a flat interval

$$H'(\xi) = v_j \left(1 + \frac{\alpha_j \eta}{\xi}\right) e^{-\alpha_j \eta/\xi} - \lambda. \tag{98}$$

If $\eta = 0$ this gives

$$H'(\xi) = v_j - \lambda. \tag{99}$$

If $\eta > 0$ we use formula (79) to obtain

$$H'(\xi) = \frac{\mu}{\alpha_j} \left(1 + \log \frac{\alpha_j v_j}{\mu}\right) - \lambda. \tag{100}$$

On a non-initial interval with λ between values of v_i we find, starting from (97) and using (95), (79), and (77), that

$$H'(\xi) = \lambda'(1 - \xi) - \lambda + \frac{\mu}{\alpha_j}\left(1 + \log\frac{\alpha_j v_j}{\mu}\right) - \alpha_j v_j e^{-\alpha_j \eta/\xi}\eta'$$

$$= \frac{\mu}{\alpha_j}\left(1 + \log\frac{\alpha_j v_j}{\mu}\right) - \lambda + \eta'\left[\frac{\lambda}{\omega(\lambda)}(1-\xi) - \alpha_j v_j e^{-\alpha_j \eta/\xi}\right]$$

$$= \frac{\mu}{\alpha_j}\left(1 + \log\frac{\alpha_j v_j}{\mu}\right) - \lambda + \eta'\left[\frac{\lambda}{\omega(\lambda)}(1-\xi) - \mu\right]$$

$$= \frac{\mu}{\alpha_j}\left(1 + \log\frac{\alpha_j v_j}{\mu}\right) - \lambda.$$

Thus formula (100) holds for all cases when $\eta > 0$, and, taken along with (99), gives the complete formula for $H'(\xi)$. In particular, $H'(\xi)$ is continuous on the interval $[0, 1)$. We need one more important fact.

Lemma 11. *If* $H'(\xi) = 0$ *at a point where* λ *is not equal to any* v_i, *and* $\lambda > \lambda^0$, *then* $H''(\xi) < 0$.

Proof. The equation $H'(\xi) = 0$ may be written, using (100) and (79), in the form

$$\frac{\mu\eta}{\xi} - \lambda = -\frac{\mu}{\alpha_j}. \tag{101}$$

Starting from formula (100), we get now

$$H''(\xi) = \frac{1}{\alpha_j} \log \frac{\alpha_j v_j}{\mu} \mu' - \lambda'$$

$$= \frac{1}{D} \left[\left(\frac{\mu\eta}{\xi} - \lambda \right) \frac{\eta}{\xi} - \frac{\lambda}{\omega(\lambda)} \left(\frac{\eta\omega}{\xi} + \frac{\lambda\xi}{\alpha_j\mu} \right) \right], \tag{102}$$

where D is given by (96). Substituting (101) into (102), we get

$$H''(\xi) = -\frac{1}{D} \left[\frac{\mu\eta}{\alpha_j\xi} + \frac{\lambda}{\omega(\lambda)} \left(\frac{\eta\omega}{\xi} + \frac{\xi\lambda}{\alpha_j\mu} \right) \right] < 0,$$

as we set out to prove.

Description of the behavior of $H(\xi)$.

We may thus give a complete description of the behavior of $H(\xi)$. There are three types of intervals to consider.

(i) *Initial interval with $\eta = 0$.* This will occur if and only if $\alpha_j v_j \omega(\lambda^0) < \lambda^0$; see Lemma 8. On such an interval $H(\xi)$ is linear and is given by

$$H(\xi) = (1 - \xi) \lambda^0 + v_j \xi. \tag{103}$$

(ii) *Interval on which $\lambda = v_i > \lambda^0$.* On such an interval $H(\xi)$ is strictly convex. It is accordingly necessary only to consider the end points. The derivative is given by (100). There is always a terminal such interval; see Lemma 9.

(iii) *Noninitial interval $[a, b]$ on which $\lambda \neq v_i$ for all i.* On such an interval, from Lemma 11, the only interior critical points are strict maxima. If there were two such interior critical points, there would have to be a minimum between them. Hence there is at most one interior critical point. If there is one, it is the maximum on that interval. If there is none, $H(\xi)$ is either strictly increasing or strictly decreasing.

These facts make it very easy to determine whether there is a maximum on such an interval of type (iii). Suppose $H'(a) > 0$. If $H'(b) \geq 0$ the maximum is at b. If $H'(b) < 0$ there is an interior maximum on $[a, b]$. If $H'(a) \leq 0$ the maximum is at a and there is no interior critical point.

Note. If we establish that the maximum is inside an interval of type (iii), then we must have $H'(\xi) = 0$ as well as (77), (78), and (79). If the value of λ is known well enough so that the constant value of $\omega = \omega(\lambda)$, and the value of

$$\theta = \sum_{v_i > \lambda} \frac{1}{\alpha_i} \log v_i \tag{104}$$

are known, one may solve (77), (78), and (79) to yield

$$\log\mu = \frac{\log\dfrac{1-\xi}{\omega} + \dfrac{\theta-1}{\omega} + \dfrac{\xi}{\alpha_j\omega}\log v_j\alpha_j}{1+\dfrac{\xi}{\alpha_j\omega}}, \tag{105}$$

$$\log\lambda = \frac{\dfrac{\xi}{\alpha_j\omega}\log\dfrac{\alpha_j v_j\omega}{1-\xi} + \dfrac{\theta-1}{\omega}}{1+\dfrac{\xi}{\alpha_j\omega}}, \tag{106}$$

and

$$\eta = \frac{\dfrac{\xi}{\alpha_j\omega}}{1+\dfrac{\xi}{\alpha_j\omega}}\left[1-\theta+\omega\log\frac{\alpha_j v_j\omega}{1-\xi}\right]. \tag{107}$$

One then has the explicit formula:

$$H'(\xi) = \frac{\omega(1-\xi)\log\dfrac{\alpha_j v_j\omega}{1-\xi} - \dfrac{\xi^2}{\alpha_j} + \xi\left(\dfrac{1}{\alpha_j}-2\omega+\theta-1\right)+\omega-\theta+1-\alpha_j\omega^2}{\dfrac{\omega\alpha_j}{\mu}(1-\xi)\left(1+\dfrac{\xi}{\alpha_j\omega}\right)}. \tag{108}$$

The numerator in (108) is concave in ξ, so that the Newton method may be used with ease for the solution of the equation $H'(\xi)=0$ in a practical case. It is also convenient to note the formula

$$H(\xi) = \mu\omega\left(1+\frac{\xi}{\alpha_j\omega}\right) \tag{109}$$

which may along with (105) be used for explicit plotting.

Two special results.

Theorem IV. *Let λ^0 be the value of the i-problem with all the resources of both sides devoted to it. Let $\omega=\omega(\lambda^0)$ correspond to it under formula (76). Then:*

1) *if all $v_j<\lambda^0$, all $x_j=0$;*
2) *if all $v_j\leq\lambda^0$ and*

$$\frac{\omega\alpha_j v_j}{\lambda^0}\geq 1 \tag{110}$$

for all the j with $v_j=\lambda^0$, all $x_j=0$;

3) *if condition (110) holds along with*

$$\frac{1}{\omega\alpha_j}\left(1+\log\frac{\alpha_j v_j\omega}{\lambda^0}\right)>1 \tag{111}$$

for some j, some $x_j>0$;

4) if some $v_j > \lambda^0$ and condition (110) does not hold for that j, then some $x_j > 0$;

5) if all $v_j \leq \lambda^0$, and for some j both $v_j = \lambda^0$ and condition (110) does not hold, there exists a solution with that x_j positive but there also exists a solution with all x_j zero.

Proof. We first prove 1). We recall as in inequality (23) that on $[\xi^*, 1]$

$$H'(\xi) = \frac{\mu}{\alpha_j}\left(1 + \log\frac{\alpha_j v_j}{\mu}\right) - \lambda < v_j - \lambda. \tag{112}$$

But λ is nondecreasing, so that $\lambda \geq \lambda^0$ on $[\xi^*, 1]$. Hence $H'(\xi) < 0$ on $[\xi^*, 1]$. If there is a nondegenerate initial interval, formula (103) shows that $H'(\xi) < 0$ there as well. Hence for all j the maximum of $H(\xi)$ is at $\xi = 0$ and 1) is proved.

As to 2), if (110) holds for all j with $v_j = \lambda^0$, then, from Lemma 8 [formula (84)], $\xi^* = 0$ and inequality (111) may again be applied to show that $H'(\xi) < 0$ throughout for each j.

As to 3), again $\xi^* = 0$. From equation (77), $\mu \to \dfrac{\lambda^0}{\omega}$ as $\xi \to 0$. Hence, from formula (103),

$$H'(\xi) \to \frac{\lambda^0}{\omega\alpha_j}\left(1 + \log\frac{\alpha_j v_j \omega}{\lambda^0}\right) - \lambda^0$$

as $\xi \to 0$. (111) then implies that $H(\xi)$ is increasing near $\xi = 0$.

As to 4), if condition (110) does not hold there is a nondegenerate initial interval $[0, \xi^*]$ on which $H'(\xi) = v_j - \lambda^0 > 0$.

As to 5), then $H(\xi)$ is constant on the nondegenerate interval $[0, \xi^*]$, from which the statement follows.

Theorem V. If

$$\underset{j}{\text{Max}}\ v_j e^{-\alpha_j}(1 + \alpha_j) \leq \underset{i}{\text{Max}}\ v_{i'}$$

some $x_i > 0$ in the solution.

Proof. Suppose the solution has $x_j = \xi = 1$. Then $H'(1) > 0$. Because of Lemma 9, $\lambda(1) = \underset{i}{\text{Max}}\ v_i$. As to μ, we use formula (79), which gives

$$\alpha_j = \log\frac{\alpha_j v_j}{\mu} \quad \text{or} \quad \mu = \alpha_j v_j e^{-\alpha_j}.$$ Using these and formula (100), we get

$$H'(1) = v_j e^{-\alpha_j}(1 + \alpha_j) - \underset{i}{\text{Max}}\ v_i > 0,$$

which contradicts our hypothesis.

Addendum

Simultaneously vulnerable systems

This addendum treats a Max-Min problem arising in practice and much like the one treated above. The reason for its inclusion is its importance in the applications, as will be seen in the concluding sentences. Put

$$G(x, z) = \sum_{i=1}^{n} v_i x_i e^{-\alpha_i y} + \sum_{j=n+1}^{m} v_j x_j e^{-\alpha_j y_j / x_j}. \tag{113}$$

Here the z denotes the vector $(y, y_{n+1}, \ldots, y_m)$ satisfying

$$y + \Sigma y_j = 1, \ y, \ y_j \geqq 0. \tag{114}$$

The x-strategy $(x_1, \ldots, x_n, x_{n+1}, \ldots, x_m)$ is always subjected to the side conditions

$$x_i \geqq 0, \quad x_j \geqq 0, \tag{115}$$

and will later be subjected to the side conditions

$$\Sigma x_i + \Sigma x_j = 1. \tag{116}$$

The explanation of the heading is as follows: in the example giving rise to the payoff (113) there are several submarine systems, of different "vulnerabilities" α_i and "effectivenesses" v_i, which are affected by the same anti-submarine effort y. In the model of the main text the systems are affected separately and independently.

Put

$$\varphi(x) = \underset{z}{\text{Min }} G(x, y), \tag{117}$$

the minimum being taken under conditions (114).

Our object is to maximize $\varphi(x)$ subject to conditions (116). By studying $\varphi(x)$ in much the same way as in the body of the chapter we are able to calculate a directional derivative in any direction γ. This again is not included in the theory of Chapter III, since the x_j in the denominators on the right side of (113) lead to the non-continuity of the derivatives of $G(x, z)$ at points where $x_j = y_j = 0$. This calculation is tedious, for reasons similar to those previously encountered, the difficulty occurring in differentiation at points where $\Sigma \alpha_i v_i x_i = \mu$, the μ corresponding to the solution for z in terms of x (see below, formula (120). The result is

$$D_\gamma \varphi(x) = \sum_{i=1}^{n} \gamma_i v_i e^{-\alpha_i y} + \mu \sum_{\alpha_j v_j > \mu} \frac{\gamma_j}{\alpha_j} \left[1 + \log \frac{\alpha_j v_j}{\mu} \right] + \\ + \sum_{\alpha_j v_j \leqq \mu} \gamma_j v_j, \tag{119}$$

where y is given by the solution to the equation

$$\sum_{i=1}^{n} \alpha_i v_i x_i e^{-\alpha_i y} = \mu \tag{120}$$

if $\Sigma \alpha_i v_i x_i > \mu$, and by zero in the contrary case. The remainder of the formula for the solution $z = (y, y_{n+1}, ..., y_m)$ in terms of x is:

$$y_j = \frac{x_j}{\alpha_j} \log \frac{\alpha_j v_j}{\mu} \tag{121}$$

if $\alpha_j v_j > \mu$ and $y_j = 0$ otherwise. The μ is determined from equation (114), and depends continuously on x.

It follows that $D_\gamma \varphi(x)$ is continuous in x. At points where μ is not equal to any of the values $\alpha_j v_j$ and where $\Sigma \alpha_i v_i x_i > \mu$, it is possible to differentiate μ with respect to x and thus $D_\gamma \varphi(x)$ as well. Write

$$A = \frac{1}{\Sigma \alpha_i^2 v_i x_i}, \tag{122}$$

$$B = \sum_{\alpha_j v_j > \mu} \frac{x_j}{\mu \alpha_j}, \tag{123}$$

$$C = A \cdot \Sigma \alpha_i v_i \gamma_i e^{-\alpha_i y}, \tag{124}$$

$$D = \sum_{\alpha_j v_j > \mu} \frac{\gamma_j}{\alpha_j} \log \frac{v_j \alpha_j}{\mu}. \tag{125}$$

Use the notation $d_\gamma y$ to mean the partial derivative of y in (120) with μ constant and x moving in the direction γ, and y_μ to mean the partial derivative of y in (120) with respect to μ. Then we have

$$d_\gamma y = -C \tag{126}$$

and

$$y_\mu = -A. \tag{127}$$

Differentiating in (114) with μ away from values of $\alpha_j v_j$ we get

$$y_\mu \mu' + d_\gamma y + \sum_{v_j \alpha_j > \mu} \frac{\gamma_j}{\alpha_j} \log \frac{\alpha_j v_j}{\mu} - \mu' \sum_{v_i \alpha_i > \mu} \frac{x_j}{\alpha_j \mu}; \tag{128}$$

here μ' denotes the total derivative of μ in the γ direction. Rewriting (128) with the aid of (126) and (127), and the formulas (122)—(125), we get

$$\mu' = \frac{D - C}{A + B}. \tag{129}$$

Now we are ready to differentiate (119) in the case $\Sigma \alpha_i x_i v_i > \mu$ and μ

not equal to a value of $\alpha_j v_j$. We get in this case

$$D_\gamma(D_\gamma \varphi(x)) = \mu' \left[-\Sigma \alpha_i v_i \gamma_i e^{-\alpha_i y} \cdot y_\mu + \sum_{\alpha_j v_j > \mu} \frac{\gamma_j}{\alpha_j} \log \frac{\alpha_j v_j}{\mu} \right]$$

$$= \frac{D^2 - C^2}{A + B}. \tag{130}$$

It is thus easy to see that $D_\gamma(D_\gamma \varphi(x)) \leqq 0$ in directions with the $\gamma_j = 0$ and $D_\gamma(D_\gamma \varphi(x)) \geqq 0$ in directions with the $\gamma_i = 0$. In the case $\Sigma \alpha_i v_i x_i < \mu$ the calculation is relatively trivial and yields the same results. At points where μ is equal to some of the $\alpha_j v_j$ or to $\Sigma \alpha_i v_i x_i$, the second derivative does not exist. However, a calculation analogous to the one yielding formula (56) for the separately vulnerable case shows that the upper limits of the "right" (i.e. in the direction y) difference quotients in the first case are non-positive and the lower limits of the left difference quotients in the second case are non-negative. Thus it follows as before that the following lemma is true:

Lemma. *For the simultaneously vulnerable problem* (113), *the function* $\varphi(x)$ *is concave in the* x_i *and convex in the* x_j.

It follows exactly as in the main text that the statement of Theorem III on the possibility that more than one x_j should be positive in the solution holds, without change in wording, for the present problem.

Thus one can solve this problem by considering one value of j at a time.

But there is a further great simplification.

Theorem VI. *There always exists a solution to the simultaneously vulnerable problem with at most one* x_i *positive.*

Proof. This proof follows the "Description of the Behavior of $H(\xi)$" in the preceding text. The reader is invited (Exercise 12) to fill in detail.

Fix ξ and solve the problem

$$H(\xi) = \underset{\Sigma x_i = 1 - \xi}{\text{Max}} \underset{y + \eta = 1}{\text{Min}} \Sigma v_i x_i e^{-\alpha_i y} + v_j \xi e^{-\alpha_j \eta / \xi}. \tag{131}$$

This in fact is a game. There therefore exist λ and μ such that

$$v_i e^{-\alpha_i y} = \lambda \quad \text{if} \quad x_i > 0,$$
$$\leqq \lambda \quad \text{if} \quad x_i = 0, \tag{132}$$

(120) once again,

$$\Sigma \alpha_i v_i x_i e^{-\alpha_i y} = \mu \quad \text{if} \quad y > 0,$$
$$\leqq \mu \quad \text{if} \quad y = 0, \tag{133}$$

and finally as before

$$\eta = \frac{\xi}{\alpha_j} \log \frac{\alpha_j v_j}{\mu} \quad \text{if} \quad \alpha_j v_j > \mu,$$
$$= 0 \quad \text{if} \quad \alpha_j v_j \leqq \mu. \tag{134}$$

These give us

$$H(\xi) = \lambda(1 - \xi) + v_j \xi e^{-\alpha_j \eta/\xi}. \tag{135}$$

Just as in the "Description" one now proves that λ and η are uniquely defined continuous functions of ξ, and μ as well when $\eta > 0$. It follows from (132) that $\lambda = \lambda(y)$, where

$$\lambda(y) = \underset{i}{\text{Max}}\; v_i e^{-\alpha_i y}. \tag{136}$$

As a function of y, $\lambda(y)$ is strictly decreasing and convex. At a corner point where two or more of the curves $v_i e^{-\alpha_i y}$ intersect the one coming in from the left is the one with the maximum α_i and the one going out to the right is the one with minimum α_i.

We may now write (133) in the form

$$\begin{aligned}
\lambda(y) \Sigma \alpha_i x_i &= \mu && \text{if}\quad y > 0, \\
&\leq \mu && \text{if}\quad y = 0.
\end{aligned} \tag{137}$$

With an argument similar to the proof of Lemma 6 one proves that η and therefore λ is nondecreasing in ξ. If $y > 0$ for points arbitrarily close to $\xi = 1$, equality would hold in (137). But the left side approaches zero, so that μ does as well, so that from (134) η approaches infinity, a contradiction. Hence there is a non-degenerate terminal interval $[\xi^{**}, 1]$ on which $\eta = 1$. On this interval $\lambda = \lambda(0) = \bar{v} = \underset{i}{\text{Max}}\; v_i$. There is an analog to Lemma 5 on initial intervals $[0, \xi^*]$.

The most important observation for our purposes is the following analog of Lemma 10. *If λ is a constant on a maximal non-degenerate subinterval other than the initial interval or the terminal interval, that constant is a corner value of λ, i.e. $v_i e^{-\alpha_i y} = v_{i'} e^{-\alpha_{i'} y}$ for some pair (i, i'). Conversely when λ passes a corner value there is such a non-degenerate interval.*

The proof of this statement follows Lemma 10. Given such a non-degenerate interval, since η is constant μ must be increasing. But y is fixed and positive, so that equality holds in (137). If there were only one term in the sum the left side would be $\alpha_i x_i \lambda(y) = \alpha_i (1 - \xi) \lambda(y)$, so that it would decrease. Hence there are at least two terms, with different values of α_i.

On the other hand, suppose that λ passes such a corner value, attaining it at a single point ξ. Then for points ξ' just to the left of ξ we have $\lambda' \,\text{Min}\,\alpha_i(1 - \xi') = \mu'$, and for points ξ'' just to the right we have $\lambda'' \,\text{Max}\,\alpha_i(1 - \xi'') = \mu''$, the minimum and maximum being taken over the curves intersecting at that corner value. On taking the limits we get $\text{Max}\,\alpha_i = \text{Min}\,\alpha_i$, a contradiction.

The proof of the theorem is now easy. First suppose that the solution is in the initial interval $[0, \xi^*]$. Then there is a solution at $\xi = 0$ or at

$\xi = \xi^*$. If $\xi = 0$ the statement of the theorem is trivial. If $\xi = \xi^*$ we observe that the interval to the right of ξ^* has η and λ strictly increasing. From (137), for points just to the right of ξ^* we have $\lambda(y)\alpha_i(1 - \xi) = \mu$, and from (134) we know that $\mu \to \alpha_j v_j$. In the case that there are several equal values of $v_i e^{-\alpha_i}$, the α_i in question is the maximum one. We thus get in the limit $\lambda(1) \operatorname*{Max}_i \alpha_i(1 - \xi^*) = \mu = \alpha_j v_j$. If then $x_i^* > 0$ for any α_i not the maximum of those corresponding to the equal values of $v_i e^{-\alpha_i}$, $\lambda(1) \Sigma \alpha_i x_i^* < \mu$, a contradiction. Hence the solution at ξ^* is $x_i^* = 1 - \xi^*$ for the i yielding $\operatorname*{Max}_i \alpha_i$, and otherwise $x_i^* = 0$, with the obvious alteration if there are two systems with identical α_i and v_i.

The solution cannot lie interior to one of the non-initial intervals corresponding to constant corner values of λ, since on these intervals η is constant as well and $H(\xi)$ is strictly convex. If the solution lies interior to an interval on which λ is varying, only one x_i can be positive because λ is not a corner value. Finally, if it lies on the edge of a flat interval, an argument just like the one used for a point on the edge of the initial interval shows that only one x_i can be positive.

The proof is complete.

This theorem in combination with the Lemma shows that we need only consider one system of type i and one system of type j at a time. Thus:

The simultaneously vulnerable problem with n systems of type i and $m - n$ systems of type j may be treated by doing $n(m - n)$ problems of the type treated in the main text with one i and one j each,

These results may be generalized somewhat. Suppose there are several classes of percentage-vulnerable systems, instead of only one as in formula (113). We suppose each class is vulnerable to the same attack, i.e. there are several terms in place of the first term of (113). *Then there will be, except for trivialities, only one representative of each class in the mix.* Thus this generalized problem reduces to the original problem (1). We omit the proof.

In the applications, one frequently encounters the following situation. There is only one class of simultaneously vulnerable systems, mounted on submarines. The submarines are identical except for the missiles, thus have the same vulnerability. Similarly the numerically vulnerable systems are equally vulnerable but have different warheads. Then it is only necessary to take the submarine system with the largest value of v_i and compare it with the numerically vulnerable system with the largest value of v_j. Using the information on $H(\xi)$ in the main text, this is quite simply done. One calculates ξ^* from (81) and ξ^{**} from (87) and (88). Then there are only two or three intervals to consider, and $H(\xi)$ is convex on the right one and linear on the interval $[0, \xi^*)$ if that is not degenerate. The calculations may therefore be done very easily by hand.

In Chapter VII, Section 8, we shall see how to reduce a problem even more complex in appearance to the simple study of $H(\xi)$ with one i and one j.

Examples

A hand calculation. Take first one system of type i and one of type j, with $v_i = v_j = \alpha_i = \alpha_j = 1$. We start by using Lemma 8 to find ξ^*. We have $\lambda^0 = v_1 e^{-\alpha_1} = e^{-1}$. We find

$$\omega(\lambda^0) = \sum_{v_i > \lambda^0} \frac{1}{\alpha_i} = 1.$$

Hence the right side of (84) is equal to

$$1 - \frac{\alpha_j v_j \omega(\lambda^0)}{\lambda^0} = 1 - e < 0.$$

It follows that $\xi^* = 0$.

We turn to Lemma 9 to calculate ξ^{**}. Equation (87) and (88) reduce on eliminating the μ to

$$e^{-1/\xi} = 1 - \xi. \tag{138}$$

We find (by slide rule) $\xi^{**} \doteq .740$. The value of $H(\xi)$ at ξ^{**} is

$$(1 - \xi^{**})\bar{v} + v_j \xi^{**} e^{-\alpha_j/\xi^{**}}$$
$$= (1 - \xi^{**}) + \xi^{**} e^{-1/\xi^{**}}$$
$$\doteq .450 > e^{-1} = v_j e^{-\alpha_j}.$$

Since $H(\xi)$ is convex on $[\xi^{**}, 1)$, this proves that the maximum is on the interval $[0, \xi^{**}]$.

We use formula (98) to calculate $H'(0)$. Since $\eta = 0$ when $\xi = 0$ we get

$$H'(0) = v_j - \lambda^0 = 1 - e^{-1} \doteq .631.$$

It follows that the maximum is interior to the interval $(0, .740)$.

Now we have four equations as follows:

$$\left(1 + \frac{\eta}{\xi}\right) e^{-\eta/\xi} = \lambda; \tag{139}$$

$$\frac{\mu}{\lambda} = 1 - \xi; \tag{140}$$

$$\log \frac{1}{\lambda} = 1 - \eta; \tag{141}$$

$$\eta = \xi \log \frac{1}{\mu}. \tag{142}$$

One may (Exercise 13) eliminate successively η, μ, and λ and get the equation in ξ alone [10] :

$$\frac{1-\xi}{1+\xi}\left(1+\log\frac{1}{1-\xi}\right)=\xi. \tag{143}$$

Equation (143) has the solution $\xi \doteq .535$. Thus the solution is $x_i = .465$, $x_j = .535$. The value of $H(\xi)$ at this point is .486.

A machine example. Consider one system of type i and two systems of type j. This example was done by machine. We take $i = 1$, $j = 2, 3$, and assume the table

$$\alpha_1 = 1/2 \quad \alpha_2 = 1 \quad \alpha_3 = 1/2$$
$$v_1 = 2 \quad\quad v_2 = 1 \quad v_3 = 2 .$$

In this case the graph of $H(\xi)$ looks like this:

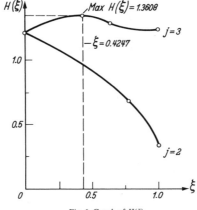

Fig. 6. Graph of $H(\xi)$

The interval $(.7412, 1)$ is an interval of constancy for η and λ for $j = 2$, and the interval $(.5791, 1)$ for $j = 3$. On these intervals $H(\xi)$ is convex, as we showed in the text. The plotting information is:

$$j = 2 \quad \xi = 0 \quad\quad H(\xi) = 1.2131$$
$$\xi = .7412 \quad H(\xi) = .7099$$
$$\xi = 1.00 \quad\quad H(\xi) = .3679$$

$$j = 3 \quad \xi = 0 \quad\quad H(\xi) = 1.2131$$
$$\xi = .4247 \quad H(\xi) = 1.3608$$
$$\xi = .5791 \quad H(\xi) = 1.3302$$
$$\xi = 1.00 \quad\quad H(\xi) = 1.2131 .$$

---

[10] Or else apply formula (108).

6*

The values of the variables at the maximum are:

$$x_1 = .5753, \quad x_2 = 0, \quad x_3 = .4247.$$

Proof that Max Min $\neq$ Min Max. Consider the simple problem

$$F(x, y) = (1 - x)e^{-(1-y)} + xe^{-y/x}.$$

We found the Max Min solution above. To find the Min Max, we note
that $F(x, y)$ is convex in x for each fixed y. Hence the solution for
Max is attained at a corner. We therefore get

$$M(y) = \operatorname*{Max}_x F(x,) = \operatorname{Max}\{e^{-(1-y)}, e^{-y}\}.$$

For $0 \le y \le 1/2$ we get $M(y) = e^{-y}$, which decreases from unity to
$e^{-1/2}$, and for $1/2 \le y \le 1$ we get $M(y) = e^{-(1-y)}$, which increases from
$e^{-1/2}$ to unity. Therefore $\operatorname*{Min}_y M(y) = e^{-1/2}$ and is assumed at $y = 1/2$.
We thus have

$$\operatorname*{Max}_x \operatorname*{Min}_y F(x, y) = .4859 < .6065 = \operatorname*{Min}_y \operatorname*{Max}_x F(x, y).$$

Hence the problem does not have a pure strategy game solution.

Exercises to Chapter V

1. Explain the passage from formula (14) to formula (15).

2. Carry through the details of the bracketed remark following (20).
Use great care.

3. Prove Lemma 1.

4. Write out in full detail the calculation of Section 2 with the
j-terms missing. This calculation should include an explanation of the
choice of the sums Σ^* and Σ^{**} in formula (47) and elsewhere.

5. Prove that formula (52), and the formula that corresponds to
it in the case $\mu^1 \ge \mu^0$, have the same limit as $x^1 \to x^0$.

6. The Schwartz inequality is: $(\Sigma a_i b_i)^2 \le \Sigma a_i^2 \Sigma b_i^2$. Prove the inequality.

7. Prove Theorem III.

8. Reduce from (65)—(69) the solution (70)—(73).

9. Prove that the problem

$$\operatorname*{Max}_{\Sigma x_i = 1 - \xi} \operatorname*{Min}_{\Sigma y_i + \eta = 1} \Sigma v_i x_i e^{-\alpha_i y_i} + v_j \xi e^{-\alpha_j \eta/\xi}$$

has a solution as a game. Deduce (70)—(73) from this observation.

10. Compute the derivatives (93), (94), (95).

11. Prove the statements made under (iii) in the "Description of the
Behavior of $H(\xi)$," using Lemma 11.

12. Fill in the details in Theorem VI.

13. Reduce the system (139)—(142) to the equation (143). After that,
try deriving (108), and check.

14. Work out the machine example (Figure 6) by hand.

Chapter VI

A model for allocation of weapons to targets

1. Introduction. This chapter is concerned with a weapons assignment problem of both theoretical and practical importance. It is a simple maximum problem, and is basic to the work of Chapter VII [11].

The *practical* problem of finding an optimal assignment of weapons of various types to targets of various types is of obvious importance. For this reason it is important to understand the limits on the application of this model. The method gives solutions in real numbers, i.e., for example, 2.31 weapons of some type allocated to some target. There is no discussion of solutions in integers. The real-number solution must therefore be regarded as a guide, and may be used as follows: one rounds it off "by eye" to an integer allocation and compares the effect of the latter with the real-number "ideal." With small weapons this approach will often be satisfactory. With large weapons the problem of, say, "rounding off" a solution that says to apply .022 thousand-megaton weapons to each of four targets, is not so easily brushed aside. In certain complex situations integer programming may well be necessary.

As to *theory*, the weapons assignment problem is central to the question of which weapons should be brought in the first place. For in considering such questions one must measure the value of a weapon in terms of the damage it does. But that value will depend on how it is used and what other weapons are available. If there are few weapons and many targets this problem often disappears, as every weapon finds its target and linear models (i.e., two weapons do twice as much damage as one) may suffice. But if there are many weapons, more than one to a target, saturation effects begin to enter. Saturation effects can quite reverse the findings of weapons selection analyses. The following is a very simple example. Suppose we are given a choice between five $1 weapons and one $5 weapon, each type intended for use against hard military targets. The weapons either destroy the target, with a certain probability, or leave it unscathed. Suppose the $1 weapon has a kill probability of .12 and the $5 weapon a kill probability of .5. If now there are five targets available the total expected

---

[11] For the understanding of Chapter VII, it is, however, not essential to read the details of this chapter beyond the point at statements about uniqueness are made.

kills from the five $1 weapons amounts to $5 \cdot .12 = .6$, and from the
$5 weapon .5. Thus with five targets the choice is for five of the $1
weapons. But suppose there were only one target. The five $1 weapons
would, if they were aimed and functioned independently, yield a kill
probability of $1 - .88^5 \doteq .472$, less than the expectation from the single
$5 weapon. Thus in the presence of saturation the choice is for one $5
weapon.

It is clear that the relative value of weapons depends on how they
are used, and what other weapons are being used. It is difficult to say
what parameter in the solution represents a "value" for an individual
weapon. For instance, the Lagrange multiplier, the marginal utility,
can have a curious reversing effect where the "better" weapon has a
lower marginal value at the optimum. The value of the *entire* collection
of weapons, rather than of the individual weapons, can be measured in
terms of the damage it does; the theoretical object of this chapter is to
provide such a measure by optimizing the allocation of those weapons.

Chapter VII will discuss the problem of choosing that mix in the first
place, using the present allocation model to assign the value.

Section 2 states the problem and finds necessary and sufficient
conditions for the solution. Section 3 states further facts about the
solution in the form of lemmas. This is the central part of this chapter.
Section 4 gives mathematically the method of finding the solution for n
weapons types given that the problem has been solved for $(n-1)$ weapons
types. It is based on the study of "linking structures".

This mathematical description of the characteristics of the solution
and the induction process for obtaining it is sufficiently detailed that
a program for machine computation should be preparable without
further analysis. As an illustration the author has solved a three-
weapons-types, five targets case by hand. This is done in three stages;
first a 1-weapon-type case is solved, then a 2-weapon-types, then the
three. This calculation is found in Sections 5 and 6. Some possible
"linking structures" are exhibited with the aid of this example.

Section 7 gives the example mentioned above of reversal of the
marginal utility. Section 8 solves a special case, when the effect of a
weapon of type i on the kth target can be written in the form

$$b_{ik} = a_i c_k .$$

This problem is strikingly easy; it is equivalent to a 1-weapon problem.
And it turns out (see Chapter VII, Section 8) to be of considerable importance.

The material that is new here is contained in Sections 3 and 4.
The author believes that the simple separable case of Section 8 has
not been published before. The technique uf Section 2 and the beginning of Section 3 are well-known.

2. Statement of the problem. Necessary and sufficient conditions.

The problem is to maximize

$$D(\xi) = \sum_{k=1}^{K} F_k(h_k), \tag{1}$$

where the F_k are arbitrary, smooth, increasing, strictly concave functions, zero at $h_k = 0$, where

$$h_k = \sum_{i=1}^{n} b_{ik}\xi_{ik} \tag{2}$$

and the ξ_{ik} are subjected to the side conditions $\xi_{ik} \geq 0$ and

$$\sum_{k=1}^{K} \xi_{ik} = 1, \quad i = 1, \ldots, n. \tag{3}$$

The ξ_{ik} denotes the *proportion* of the supply of the ith weapon which is allocated to the kth target. The effectiveness coefficient b_{ik} takes into account the total amount of the ith weapon available, so that we could write a 1 on the right side of (3). It will be the unit effectiveness of the ith weapon against the kth target, multiplied by the total supply of that weapon.

An example of $F_k(h_k)$ is the following:

$$F_k(h_k) = v_k(1 - e^{-h_k}). \tag{4}$$

Another is
$$F_k(h_k) = v_k[1 - (1 + \sqrt{h_k})e^{\sqrt{h_k}}]. \tag{5}$$

Formula (5) is of particular interest for small nuclear warheads against large area targets. Formula (4) can be applied to the case of "black and white" targets such as missile silos. Let us show how formula (4) arises. "Black and white" applied to a target means that one missile either destroys it or does not affect it. The kill probability if m weapons with independent guidance and reliability are fired at it will be, according to the usual formula,

$$1 - (1 - P_1)\cdots(1 - P_m). \tag{6}$$

If we write
$$1 - P_1 = e^{-\beta_1}, \ldots, 1 - P_m = e^{-\beta_m},$$

we get instead of (6) the expression

$$1 - e^{-\beta_1 - \cdots - \beta_m}. \tag{7}$$

If instead there were n_i weapons of type i, $i = 1, \ldots, m$, formula (7) is replaced by
$$1 - e^{-\beta_1 n_1 - \cdots - \beta_m n_m}. \tag{8}$$

Next, write the n_i as a proportion of the total supply of the ith weapon:

$$n_1 = \xi_1 x_1, \ldots, n_m = \xi_m x_m$$

where x_i denotes the total supply of the weapon of type i available for all targets including this one, and ξ_i denotes the proportion of that supply devoted to this target. Then we get instead of (8) the formula

$$1 - e^{-\beta_1 x_1 \xi_1 - \cdots - \beta_m x_m \xi_m}. \tag{9}$$

On writing $\beta_1 x_1 = b_1, \ldots, \beta_m x_m = b_m$, we get

$$1 - e^{-b_1 \xi_1 - \cdots - b_m \xi_m}. \tag{10}$$

Thus we have a formula of type (4) with v_k the value of the target and h_k given by a formula of type (2). On passing from integers to real numbers we thus have a special case of (1) subject to (3).

There is a key assumption in the formulation (1)—(2); there is at each target a "total weight of attack" h_k given by (2), and the damage $F_k(h_k)$ depends on h_k and not directly on the individual values of the $\xi_{1k}, \ldots, \xi_{nk}$. As we saw above, with black and white targets this assumption can be justified; with others it may not be.

We now state conditions necessary for the solution. We use the Gibbs lemma, applying it separately to each of the n side conditions (3). We deduce for any solution matrix $\xi = \|\xi_{ik}\|$ the existence of $\lambda_1, \ldots, \lambda_n$ such that

$$\begin{aligned} F'_k(h_k)b_{ik} &= \lambda_i \quad \text{if} \quad \xi_{ik} > 0, \\ &\leq \lambda_i \quad \text{if} \quad \xi_{ik} = 0. \end{aligned} \tag{11}$$

These conditions are necessary conditions on the solution. It happens that they are also sufficient. If there exist sets of numbers $\lambda_1, \ldots, \lambda_n$; $h_1, \ldots, h_K$, and a matrix ξ satisfying (2), (3) and (11), then ξ is a solution. We shall use this fact in finding the solution. The proof is left as an exercise for the reader (Exercise 1).

3. Some characteristics of the solution.   Here we state some facts and make some definitions essential for the construction of the solution.

First, the $h_1, \ldots, h_K$ are unique. Suppose given two solutions ξ and ξ' with corresponding sets h and h'. Put $\xi'' = \dfrac{\xi + \xi'}{2}$. Then $h'' = \dfrac{h + h'}{2}$. If $h \neq h'$, $h_k \neq h'_k$ for some k and, from the strict concavity of F_k,

$$F_k(h''_k) > \frac{F_k(h_k) + F_k(h'_k)}{2}. \tag{12}$$

Since, for the remaining values of k, (12) holds with $>$ replaced by $\geq$, then

$$D(\xi'') > \frac{D(\xi) + D(\xi')}{2},$$

a contradiction since both ξ and ξ' yield the maximum. Hence $h_k = h'_k$ for all k.

Secondly, the $\lambda_1, \ldots, \lambda_n$ are unique. Let ξ' be a second solution. We have proved that $h' = h$. For each i, there is a k_i such that $\xi_{ik_i} > 0$. Hence from (11)

$$\lambda_i = F'_{k_i}(h_{k_i})b_{ik_i} = F'_{k_i}(h'_{k_i})b_{ik_i} \leq \lambda'_i.$$

Hence $\lambda_i \leq \lambda'_i$ for all i. Since the reverse inequality also obviously holds, $\lambda_i = \lambda'_i$ for all i.

It is however easy to see that in general the matrix ξ is not unique. Suppose for instance that all the functions F_k were the same, and all the $b_{ik} = 1$. Then evidently labels may be switched without changing the problem. Since the sets $\lambda_1, \ldots, \lambda_n$ and $h_1, \ldots, h_K$ are uniquely defined by the problem, then $\lambda_1 = \cdots = \lambda_n = \lambda$ and $h_1 = \cdots = h_K = h$. All the expressions $F'_k(h_k)b_{ik}$ are equal to each other and hence $F'_k(h_k)b_{ik} = \lambda_i$, i.e., $F'(h) = \lambda$. It follows from sufficiency that any matrix ξ with $\xi_{ik} \geq 0$ satisfying

$$\Sigma_k \xi_{ik} = 1, \quad i = 1, \ldots, n$$

and

$$\Sigma_i \xi_{ik} = h, \quad i = 1, \ldots, K,$$

is a solution. But here are $n + K$ equations in nK unknowns, and the solution cannot be unique if $n > 2$ or $K > 2$.

This non-unique character of the matrix ξ constitutes a main difficulty in finding the solution. We should remark that we are not trying to find all solutions, but rather just one.

We now make some basic definitions. If, for $i \neq i'$

$$\frac{F'_k(h_k)b_{ik}}{\lambda_i} = \frac{F'_k(h_k)b_{i'k}}{\lambda_{i'}} = 1, \tag{13}$$

we say that the pairs (i, k) and (i', k) are *linked*. We denote the *link* by $[(i, k), (i', k)]$. We define a *chain* to be a succession of links

$$[(i_1, k_1), (i_2, k_1)], \quad [(i_2, k_2), (i_3, k_2)], \ldots, [(i_{m-1}, k_{m-1}), (i_m, k_{m-1})] \tag{14}$$

with $k_s \neq k_{s'}$ if $s \neq s'$.

If there is a chain with $(i, k) = (i_1, k_1)$ and $(i', k') = (i_m, k_{m-1})$, the pair (i, k) is said to be *chained* to the pair (i', k'). The relationships thus set up are referred to as a "linking structure."

Examples of linking structures are given in Section 6, matrices (53), (54) and (55).

The relation: (i, k) is chained to (i', k'), is evidently symmetric. It is not reflexive. If an element were chained to itself, the first and last link would occupy the same column, i.e., $k_{m-1} = k_1$, which is contrary to the definition. As to transitivity, suppose for instance in the case $n = 3$, $K = 2$, that the equation

$$\frac{F'_k(h_k)b_{ik}}{\lambda_i} = 1 \tag{15}$$

holds for the pairs $(1, 1)$, $(2, 1)$, $(2, 2)$, and $(3, 2)$. Then the links are $[(1, 1), (2, 1)]$ and $[(2, 2), (3, 2)]$ and their reverses. The elements $(1, 1)$ and $(3, 2)$ are chained by $[(1, 1), (2, 1)]$, $[(2, 2), (3, 2)]$; $(3, 2)$ and $(2, 2)$ are chained by $[(3, 2), (2, 2)]$, but $(1, 1)$ is not chained to $(2, 2)$. Hence the relation is not transitive. However, the following statement holds.

Lemma 1. *Suppose that*
1) (i, k) *is chained to* (i', k');
2) *there is an* (i', k^*) *chained to* (i'', k'');
3) $i \neq i''$.
Then there is an $(i'', \bar{k})$ *chained to* (i, k).

Proof. Suppose that $(i, k) = (i_1, k_1)$ and $(i', k') = (i_m, k_{m-1})$ in the chain (14), and that $(i', k^*) = (i'_1, k'_1)$ and $(i'', k'') = (i'_{m'}, k'_{m'-1})$ in a chain (14'). If the columns in the two chains are distinct, the second may be adjoined to the first to form a new chain, i.e.,

$$[(i_1, k_1), (i_2, k_1)], \ \ldots, \ [(i_{m-1}, k_{m-1}), (i_m, k_{m-1})], \ [(i'_1, k'_1), (i'_2, k'_1)],$$
$$\ldots, \ [(i'_{m'-1}, k'_{m'-1}), (i'_{m'}, k'_{m'-1})],$$

the joining being possible since $i_m = i' = i'_1$. This chain chains $(i, k) = (i_1, k_1)$ to $(i'', k'') = (i'_{m'}, k'_{m'-1})$.

Now suppose $k_s = k'_{s'}$ for some pair s, s', the s being taken as small as possible. If $i_s = i''$, the chain

$$[(i_1, k_1), (i_2, k_1)], \ \ldots, \ [(i_{s-1}, k_{s-1}), (i_s, k_{s-1})]$$

already links $(i, k) = (i_1, k_1)$ to the element (i'', k_{s-1}) of the ith row. Suppose $i_s \neq i''$. There are two cases. If $i_s = i'_{s'+1}$, then $i'_{s'+1} \neq i''$, so that $m' > s' + 1$, i.e., there is a link after $[(i'_{s'}, k'_{s'}), (i'_{s'+1}, k'_{s'})]$. We now form a chain from the first $s - 1$ links (if there are any) of (14) and the last $m' - s' - 1$ of (14') as follows:

$$[(i_1, k_1), (i_2, k_1)], \ \ldots, \ [(i_{s-1}, k_{s-1}), (i_s, k_{s-1})],$$
$$[(i'_{s'+1}, k'_{s'+1}), (i'_{s'+2}, k'_{s'+1})], \ \ldots, \ [(i'_{m'-1}, k'_{m'-1}), (i'_{m'}, k'_{m'-1})].$$

Since $i_s = i'_{s'+1}$ we have thus been able to pass from the link preceding the sth in the first chain to the link following the s'th in the second; the resulting chain chains $(i, k) = (i_1, k_1)$ to the element $(i'_m, k'_{n-1}) = (i'', k'')$.

The second case has $i_s \neq i'_{s'+1}$. But then we may form a new link $[(i_s, k_s), (i'_{s'+1}, k'_{s'})]$ from the links $[(i_s, k_s), (i_{s+1}, k_s)]$ and $[(i'_{s'}, k'_{s'}), (i'_{s'+1}, k'_{s'})]$ and form the chain

$$[(i_1, k_1), (i_2, k_1)], \ \ldots, \ [(i_{s-1}, k_{s-1}), (i_s, k_{s-1})],$$
$$[(i_s, k_s), (i'_{s'+1}, k'_{s'})], \ [(i'_{s'+1}, k'_{s'+1}), (i'_{s'+2}, k'_{s'+1})],$$
$$\ldots, \ [(i'_{m'-1}, k'_{m'-1}), (i'_{m'}, k'_{m'-1})]$$

from the first $s - 1$ links of (14), the new link, and the last $m' - s' - 1$ links of (14'). This chain links $(i, k) = (i_1, k_1)$ to the element $(i'_{m'}, k'_{m'-1}) = (i'', k_{m'-1})$ and the lemma is proved.

Transfer elements are defined for the purposes of the induction step in the next section. A pair (i, k) with $i < n$ is a *transfer element* if it is chained to an element of the nth row.

Lemma 2. *Suppose a pair* (i, k) *with* $i < n$ *is chained to a pair which lies in the same row as a transfer element. Then* (i, k) *is a transfer element.*

This lemma is an immediate consequence of Lemma 1. We now need some statements about uniqueness.

A *linking set* J_k is the set of all pairs (i, k) satisfying (15) for the given k.

Lemma 3. *Suppose given two solutions* ξ *and* ξ'. *Put* $\delta_{ik} = \xi_{ik} - \xi'_{ik}$. *Then, for any* k,

$$\sum_{i=1}^{n} \lambda_i \delta_{ik} = 0. \qquad (16)$$

Proof. It is evidently sufficient to prove (16) for i restricted to the linking set J_k. Since h_k is unique,

$$h_k = \Sigma_{J_k} b_{ik} \xi_{ik} = \Sigma_{J_k} b_{ik} \xi'_{ik},$$

so that $$\Sigma_{J_k} b_{ik} \delta_{ik} = 0. \qquad (17)$$

But from (18) $b_{ik} = \lambda_i / F'_k(h_k)$, so that

and the lemma is proved. $$\Sigma_{J_k} \lambda_i \delta_{ik} = 0$$

Lemma 4. *Let* Z_i *be the set of non-transfer elements in the* i*th row. Then the sum*

$$\sum_{k \in Z_i} \xi_{ik}$$

is unique.

Proof. Let Z denote the set of all pairs (j, k) which appear as left or right elements of links of chains starting at elements of Z_i. From Lemma 2, it follows that any element of the ith row arrived at in this way was already in Z_i[12]. The set Z now arranges itself as a union of linking sets:

$$Z = \bigcup_{k \in S} J_k.$$

Now suppose there are two solutions ξ and ξ'. Denote by $Z_j, j = 1, \ldots, n$, the set of pairs $(j, k) \in Z$ with a given value of j. Evidently Z_n is empty. Write

$$\delta_j = \sum_{k \in Z_j} \xi_{jk} - \sum_{k \in Z_j} \xi'_{jk},$$

$j = 1, \ldots, n$. Using now Lemma 3, we find that

$$\sum_{j=1}^{n} \lambda_j \delta_j = \bigcup_{k \in S} \sum_{j=1}^{n} \lambda_j \delta_{jk} = 0. \qquad (18)$$

There are now two possibilities. First, there may be no linked non-transfer elements in Z_i. Then either $F'_k(h_k) b_{ik} < \lambda_i$ and both ξ_{ik} and

---

[12] This crucial step is the reason for the suppression of column-repetitions in the definition of chains and the consequent somewhat complicated argument of Lemma 1. If column-repetitions were allowed one could, for example, start up a link joining a non-transfer element to a row other than the nth row, then come back on the same link and thence through the chain beginning at a transfer element to the nth row. Thus "false" transfer elements would have been introduced.

ξ'_{ik} are zero, or else $F'_k(h_k)b_{ik} = \lambda_i$ and ξ_{ik} and ξ'_{ik} are equal because

$$b_{ik}\xi_{ik} = b_{ik}\xi'_{ik} = h_k.$$

Or else some set Z_j with $j \neq i$ is not empty. Suppose now some $\xi_{jk} > 0$. Then either the pair (j, k) is in Z_j or else it is not linked to anything (Exercise 2). In the latter case ξ_{jk} is, as in the case of non-linked elements of Z_i, determined uniquely. Thus if $\xi_{jk} \neq \xi'_{jk}$ the pair $(j, k) \in Z_j$. Hence $\delta_j = 0$. But then (18) reduces to $\lambda_i \delta_i = 0$, i.e., $\delta_i = 0$. The lemma is proved.

The ith row is said to be *admissible* if $i = n$ or

$$\sum_{k \in Z_i} \xi_{ik} < 1. \tag{19}$$

That this definition makes sense is assured by Lemma 4. A link $[(i, k), (i', k)]$ is *admissible* if both the ith row and the i'th row are admissible; otherwise it is inadmissible.

Examples of admissible linking structures are given by matrices (53), (54), and (55) with the aid of (52).

One thus obtains an admissible linking structure with admissible chains and admissible transfer elements. A row is *eligible* if it contains an admissible transfer element or if it is the nth row. Evidently an eligible row is admissible, but the converse may be false.

Given the linking set J_k, we form the subset J_k^* of those elements of J_k which are in eligible rows.

The following lemma follows from Lemma 2 applied to the admissible linking structure.

Lemma 5. *Suppose J_k^* is not empty. If $(i, k) \in J_k$ and $\xi_{ik} > 0$, then $(i, k) \in J_k^*$.*

4. The inductive method for finding the solution. We suppose given a solution for $n - 1$ weapons systems, i.e., sets $\lambda_1, ..., \lambda_{n-1}$; $h_1, ..., h_K$ and a matrix $\|\xi_{ik}\|$, $i = 1, ..., n - 1$, $k = 1, ..., K$ satisfying

$$\Sigma_k \xi_{ik} = 1, \quad i = 1, ..., n - 1, \tag{20}$$

$$\sum_{i=1}^{n-1} b_{ik}\xi_{ik} = h_k, \quad k = 1, ..., K, \tag{21}$$

and such that the entries of the matrix

$$\left\| \frac{F'_h(h_k)b_{ik}}{\lambda_i} \right\| \quad i = 1, ..., n - 1; \quad k = 1, ..., K \tag{22}$$

are less than or equal to unity, equality holding always when $\xi_{ik} > 0$.

The object is to produce similar sets when an nth weapon system is added.

We begin by writing

$$\lambda_n = \operatorname*{Max}_k F'(h_k)b_{nk}, \tag{23}$$

the h_k being those given in (21). We then set up the matrix (22) with the additional row. Everything is now satisfied except that all the $\xi_{nk} = 0$.

We shall now start lowering the λ_i for all the eligible rows *proportionately*, i.e.,

$$\lambda_i(t) = t\lambda_i$$

for i eligible, $t \leq 1$ and t near 1.

Fix attention on a linking set J_k and its subset J_k^*. If J_k^* is empty the values of ξ_{ik} will not be altered and the linking set J_k will remain as it was. If J_k^* is not empty we solve, for any $i \in J_k^*$, the equation

$$F_k'(h_k)b_{ik} = t\lambda_i \tag{24}$$

for $h_k = h_k(t)$. It does not matter which $i \in J_k^*$ was chosen.

Lemma 5 assures that with $J_k^* \neq 0$, $\xi_{ik} > 0 \Rightarrow (i, k) \in J_k^*$. Hence at the outset ($t = 1$) we have, if $J_k^* \neq 0$:

$$h_k = \sum_{i \in J_k^*} b_{ik}\xi_{ik}. \tag{25}$$

We apportion the increase from h_k to $h_k(t)$ as t begins to decrease among the elements of J_k^* as follows. Put

$$\xi_{ik}^*(t) = \xi_{ik} + \varrho_k(t), \tag{26}$$

where $\varrho_k(t)$ is chosen so that

$$\varrho_k(t) \sum_{i \in J_k^*} b_{ik} = h_k(t) - h_k. \tag{27}$$

(An additive instead of multiplicative increase $\varrho_k(t)$ was chosen since possibly the original value $h_k = 0$). $\varrho_k(t)$ increases strictly as t decreases. If the pair (i, k) is not in a J_k^* then $\xi_{ik}^*(t) = \xi_{ik}$.

Nothing changes (as we shall see) in the admissible linking structure as the lowering begins. However the quantities $\xi_{ik}^*(t)$ defined above, for $(i, k) \in J_k^*$, are increasing strictly. Thus there will be an excess in the eligible rows. Put

$$e_i(t) = \sum_{k=1}^{K} \xi_{ik}^*(t) - 1 \tag{28}$$

for an eligible row $i < n$. Since every eligible row contains an admissible transfer element, which is in some J_k^*, the $e_i(t)$ for such a row is strictly increasing from an initial value of zero.

We are going to pass this excess through to the nth row, where, of course, for t near 1 $\Sigma \xi_{nk}^*(t)$ is near zero. In doing so we shall not alter the values of $h_k(t)$.

Fix attention on an eligible row, say the ith. This row contains an admissible transfer element which we shall denote by $(i, k_1) = (i_1, k_1)$, chained to an element $(i_m, k_{m-1}) = (n, k_{m-1})$ in the nth row by a chain

(14). In each of the rows $i_1, i_2, \ldots, i_{m-1}$ through which this chain passes (which are of course all admissible) the sum taken over the non-transfer elements satisfies

$$\sum_{k \in Z_{i_s}} \xi_{i_s k} < 1 \tag{29}$$

at the outset, and hence this inequality persists for the $\xi_{ik}^*(t)$ for t close to unity, by continuity.

Observe therefore in particular that the excess in the ith row may be taken entirely from the transfer elements, i.e., in the ith row

$$\sum_{k \in Z_i} \xi_{ik}^*(t) < 1 < \sum_{k=1}^{K} \xi_{ik}^*(t).$$

Hence we may solve the equation

$$\theta_i(t) \sum_{k \in T_i} \xi_{ik}^*(t) + \sum_{k \in Z_i} \xi_{ik}^*(t) = 1 \tag{30}$$

to obtain a continuous function $\theta_i(t)$ satisfying $0 < \theta_i(t) < 1$ for $t < 1$ and t close to 1, and evidently monotone decreasing. Here T_i denotes the set of transfer elements in the ith row.

Now consider a small quantity $\sigma < 0$ of the excess in the ith row. If we reduce the value of $\xi_{i_1 k_1}^*(t)$ by the amount σ (we are holding t fixed in this discussion), we may make up for it by increasing the value of $\xi_{i_2 k_1}^*(t)$ by the amount $\sigma \lambda_{i_1}/\lambda_{i_2}$. In fact,

$$-b_{i_1 k_1} \sigma + b_{i_2 k_1} \sigma \cdot (\lambda_{i_1}/\lambda_{i_2}) = 0$$

because the elements (i_1, k_1) and (i_2, k_1) are linked. Next one decreases the value of $\xi_{i_2 k_2}^*(t)$ by the amount $\sigma \lambda_{i_1}/\lambda_{i_2}$, and makes up for it by increasing the value of $\xi_{i_3 k_2}^*(t)$ by $(\sigma \lambda_{i_1}/\lambda_{i_2}) \cdot (\lambda_{i_2}/\lambda_{i_3}) = \sigma \lambda_{i_1}/\lambda_{i_3}$. And so forth. The final result is the following:

1) The sum $\Sigma_k \xi_{ik}^*$ has gone down by σ;
2) The sum $\displaystyle\sum_{k \in Z_{i_s}} \xi_{i_s k}^*$ for $s \geq 2$ has gone up by at most $\sigma \lambda_{i_1}/\lambda_{i_s}$;
3) The sums $\Sigma_k \xi_{i_s k}^*$ have been preserved, $s \geq 2$;
4) The sum $\Sigma_k \xi_{nk}^*$ has gone up by $\sigma \lambda_{i_1}/\lambda_n$;
5) The h_k's have not changed.

Because $e_i(t)$ is small for t near unity, it is therefore possible to pass the entire excess $e_i(t)$ into the nth row. In fact, one may pass [see formula (30)] the quantity $[1 - \theta(t)] \xi_{ik}^*(t)$ for each admissible transfer element, thus replacing $\xi_{ik}^*(t)$ by $\theta_i(t) \xi_{ik}^*(t)$. We do this for all eligible i. As a result:

1) The sums $\Sigma_k \xi_{ik}^*$ have now all been replaced by unity [formula (30)].
2) The sums $\displaystyle\sum_{k \in Z_i} \xi_{ik}^*$ for eligible rows have gone up by at most the

quantity $\displaystyle\sum_{j \neq i}{}' e_j(t) \lambda_j/\lambda_i$,

the prime denoting summation over eligible rows only.

3) The sum $\Sigma \xi_{nk}^*$ has been replaced by

$$\Sigma \xi_{nk}^* + \sum_{i<n}' e_i(t) \lambda_i / \lambda_n, \tag{31}$$

the prime again referring to eligible rows.

4) $h_k(t)$ has not been changed.

Evidently this process preserves all the desired relations between the matrix $\left\| \dfrac{F_k'(h_k) b_{ik}}{\lambda_i} \right\|$, the sets $\{\lambda_i\}$, $\{h_k\}$, and the elements ξ_{ik} for i in an eligible row. What of the other elements?

Fix attention on a linking set J_k. Two things can happen.

1) J_k *intersects an eligible row.* Thus, according to Lemma 5, $\xi_{ik} > 0$ only if i is an eligible row. In this case the desired relations are maintained by the process described above. ξ_{ik} remains zero if i is not eligible, and $\dfrac{F_k'(h_k) b_{ik}}{\lambda_i}$ may well decrease if h_k is increasing.

2) J_k *does not intersect any eligible row.* Then h_k and the λ_i's remain fixed during the lowering and the quantities $F'(h_k) b_{ik} / \lambda_i$ thus also remain fixed for $i < n$.

Thus the description of the lowering process is complete. One lowers a bit, solves for the h_k and hence the excesses $e_i(t)$, and transfers that into the bottom row with the "exchange rate" λ_i / λ_n. There is evidently no change, for a while, in the linking structure or the definition of admissible rows. One continues until one of three things happens:

(i) One of the sums $\displaystyle\sum_{k \in Z_i} \xi_{ik}$, for i an eligible row, reaches unity;

(ii) A quantity $F_k'(h) b_{ik} / \lambda_i$, for i an eligible row where h_k was previously constant and J_k did not previously intersect any eligible row, reaches unity;

(iii) The sum $\displaystyle\sum_{k=1}^{K} \xi_{nk}$ reaches unity.

In cases (i) and (ii), which could happen simultaneously, one stops, redefines the linking structure, and the admissible linking structure, determines the eligible rows, and proceeds. In case (iii) the process is complete.

It can be proved (Exercise 5) that the lowering process for a given row, and therefore the entire technique, will take only a finite number of steps to reach the solution.

5. A hand-computed example for two weapons types. We take

$$F_k(h_k) = 1 - e^{-h_k}$$

and the matrix $\|b_{ik}\|$ to be

$$\|b_{ik}\| = \left\| \begin{matrix} 1 & 2 & 3 & 4 & 5 \\ 5 & 4 & 3 & 2 & 1 \end{matrix} \right\|$$

We start by solving the top problem alone, starting by lowering λ_1 from $\lambda_1 = 5$. We easily find that $\lambda_1 = 1.35$ solves the top problem. The solution matrix is now

$$\left\| \begin{matrix} 0 & .198 & .267 & .272 & .263 \\ 0 & 0 & 0 & 0 & 0 \end{matrix} \right\|,$$

with

$$e^{-h_1} = 1, \quad e^{-h_2} = .673, \quad e^{-h_3} = .449, \quad e^{-h_4} = .336, \quad e^{-h_5} = .269.$$

The appearance of the matrix of the $F'(h_k)b_{ik} = e^{-h_k}b_{ik}$ is now

$$\left\| \begin{matrix} 1 & 1.35 & 1.35 & 1.35 & 1.35 \\ 5 & 2.69 & 1.35 & .673 & .269 \end{matrix} \right\|.$$

We begin the lowering with $\lambda_2 = 5$. We have

$$\left\| \frac{F'(h_k)b_{ik}}{\lambda_i} \right\| = \left\| \begin{matrix} .746 & 1 & 1 & 1 & 1 \\ 1 & .538 & .269 & .135 & .054 \end{matrix} \right\|.$$

There are no links. Hence we may lower λ_2 alone, using the formula

$$\xi_{21} = \frac{1}{5} \log \frac{5}{\lambda_2}$$

for ξ_{21}. The stop evidently comes when (the h_2 still given by the one-weapon solution)

$$\frac{b_{22}e^{-h_2}}{\lambda_2} = \frac{2.69}{\lambda_2}$$

reaches unity, i.e., $\lambda_2 = 2\lambda_1 = 2.69$. This is Case ii of the description at the end of Section 4.

At this point we have the solution matrix

$$\left\| \begin{matrix} 0 & .198 & .267 & .272 & .263 \\ .126 & 0 & 0 & 0 & 0 \end{matrix} \right\|.$$

The other matrix is

$$\left\| \frac{F'(h_k)b_{ik}}{\lambda_i} \right\| = \left\| \begin{matrix} .4 & 1 & 1 & 1 & 1 \\ 1 & 1 & .5 & .25 & .100 \end{matrix} \right\|.$$

Now we have a link in the second column. It is admissible since $.268 + .272 + .263 < 1$. We begin now to lower λ_1 and λ_2 simultaneously, with $\lambda_2/\lambda_1 = 2$. We pass the excess in the top row through the link to the term ξ_{22}. What can happen?

There are two possibilities. First, the sum of the last three terms in the top row could reach unity (Case i) or the sum of the first two terms in

the second row could reach unity (Case iii). No new terms may appear
because the ratios

$$\left(\frac{b_{1k}e^{-h_k}}{\lambda_1}\right)\Bigg/\left(\frac{b_{2k}e^{-h_k}}{\lambda_2}\right) = \frac{b_{1k}\lambda_2}{b_{2k}\lambda_1} \tag{32}$$

remain constant during the lowering and because in each column one
of the expressions $b_{ik}e^{-h_k}/\lambda_i$ is equal to unity; the other one therefore
is also constant.

The value of λ_1 yielding the first possibility, i.e. case i, solving

$$\frac{1}{3}\log\frac{3}{\lambda_1} + \frac{1}{4}\log\frac{4}{\lambda_1} + \frac{1}{5}\log\frac{5}{\lambda_1} = 1, \tag{33}$$

is $\lambda_1 = 1.05^-$. The corresponding value of $\lambda_2 = 2.09$. During the lowering
a portion of the quantity

$$h_2 = \log\frac{2}{\lambda} \tag{34}$$

will have been passed, with a coefficient of 1/6 [see formulas (26) and
(27)] into the top row, and the same amount into the bottom row.
Thereupon the excess e_1 in the first row is passed to the bottom row
with the added coefficient $1/2 = \lambda_1/\lambda_2$. Since at $\lambda_1 = 1.05^-$ the last three
terms sum to unity, all of h_2 must by then have passed to the bottom
row, i.e.

$$\xi_{22} = \frac{1}{4}\log\frac{2}{1.05} = .162. \tag{35}$$

Since we have for the "free" term $\xi_{21} = .174$, $\xi_2 + \xi_2 = .336 < 1$ and the
first possibility occurs first, i.e. $\xi_{13} + \xi_{14} + \xi_{15} = 1$.

The solution is now

$$\begin{Vmatrix} 0 & 0 & .352 & .336 & .313 \\ .174 & .162 & 0 & 0 & 0 \end{Vmatrix}, \tag{36}$$

and [because of (32) above]

$$\begin{Vmatrix} \dfrac{F'(h_k)b_{ik}}{\lambda_i} \end{Vmatrix} = \begin{Vmatrix} .4 & 1 & 1 & 1 & 1 \\ 1 & 1 & .5 & .25 & .100 \end{Vmatrix}. \tag{37}$$

But now the link in the second column is not admissible. Hence we
leave λ_1 alone and start lowering λ_2 alone. Evidently the next stop
occurs when $b_{23}e^{-h_3}/\lambda_2 = 1$ [h_3 not having changed during this lowering (Case ii)]. Since, on inspection of (37) above, $b_{13}e^{-h_3}/\lambda_1 = 1$ when
$\lambda_1 = 1.05^-$, this will occur when $\lambda_2 = 1.05^-$. The solution matrix is now[13]

$$\begin{Vmatrix} 0 & 0 & .352 & .336 & .313 \\ .312 & .335 & 0 & 0 & 0 \end{Vmatrix},$$

---

[13] This calculation was done to three decimal places. Sometimes, therefore, the rows
in the solution matrix will be off somewhat in the third decimal place.

7 Danskin, The Theory of Max-Min

the first two entries in the second row being calculated from the formulas $\xi_{21} = \frac{1}{5} \log \frac{5}{1.05} = .313$, $\xi_{22} = \frac{1}{4} \log \frac{4}{1.05} = .336$, and

$$\left\| \frac{F'(h_k)b_{ik}}{\lambda_i} \right\| = \left\| \begin{matrix} .2 & .5 & 1 & 1 & 1 \\ 1 & 1 & 1 & .5 & .2 \end{matrix} \right\|.$$

Thus now there is a link in the third column. That link is admissible, since $\xi_{14} + \xi_{15} = .649 < 1$. We now lower λ_2 and λ_1, with $\lambda_1 = \lambda_2 = \lambda$, starting at $\lambda_1 = \lambda_2 = \lambda = 1.05$.

This lowering may stop in two possible ways. Either the sum $\xi_{14} + \xi_{15}$ reaches unity or else the sum

$$\xi_{21} + \xi_{22} + \xi_{23} = \frac{1}{3} \log \frac{3}{\lambda} + \frac{1}{2} \log \frac{4}{\lambda} + \frac{2}{5} \log \frac{5}{\lambda} - 1, \qquad (38)$$

that is, the sum of the "free" terms in the second row

$$\xi_{21} + \xi_{22} = \frac{1}{5} \log \frac{5}{\lambda} + \frac{1}{4} \log \frac{4}{\lambda},$$

the "excess" in the first row

$$\xi_{13} + \xi_{14} + \xi_{15} - 1 = \frac{1}{6} \log \frac{3}{\lambda} + \frac{1}{4} \log \frac{4}{\lambda} + \frac{1}{5} \log \frac{5}{\lambda} - 1,$$

which is transferred at the exchange rate $\lambda_1/\lambda_2 = 1$, and the accumulation of the original apportionment of $\frac{1}{6} \log \frac{3}{\lambda}$ of h_3 to the linked term in the second row, may reach unity. But since

$$\xi_{14} + \xi_{15} = \frac{1}{4} \log \frac{4}{\lambda} + \frac{1}{5} \log \frac{5}{\lambda} = \xi_{21} + \xi_{22}$$

and since $\xi_{23} > 0$ the sum (38) exceeds $\xi_{14} + \xi_{15}$ and reaches unity first. This, it is easily calculated, occurs at $\lambda_1 = \lambda_2 = .786$. At this point the solution matrix

$$\left\| \begin{matrix} 0 & 0 & .223 & .407 & .370 \\ .370 & .407 & .223 & 0 & 0 \end{matrix} \right\|, \qquad (39)$$

the entries in the third column having been determined as follows. $h_3 = \log \frac{3}{.786} = 1.3.39$; this quantity, divided by 3, is allocated between the first and second row, the excess always being transferred to the second. Since $1 - .407 - .370 = .223$, the amount .223 will appear in the top row and $1.340/3 - .223 = .223$ will appear in the bottom row. The four remaining terms are "free" and uniquely determined. Finally, again because of (32), we still have

$$\left\|\frac{F'_k(h_k)b_{ik}}{\lambda_i}\right\| = \left\|\begin{matrix} .2 & .5 & 1 & 1 & 1 \\ 1 & 1 & 1 & .5 & .2 \end{matrix}\right\|.$$

The terminal parameters are $\lambda_1 = \lambda_2 = .786$ and

$$\|h_1, h_2, h_3, h_4, h_5\| = \|1.850, 1.627, 1.339, 1.627, 1.850\|. \tag{40}$$

This completes the calculation.

6. A hand-computed example for three weapons types. Examples of linking structures.

We consider the problem of Section 5 with an additional weapon type, as follows:

$$b_{ik} = \left\|\begin{matrix} 1 & 2 & 3 & 4 & 5 \\ 5 & 4 & 3 & 2 & 1 \\ 3 & 4 & 5 & 4 & 3 \end{matrix}\right\|.$$

Using the induction procedure, we begin with the solution (39) for two weapons just obtained, and form the matrix

$$\left\|\frac{F'_k(h_k)b_{ik}}{\lambda_i}\right\| = \left\|\frac{e^{-h_k}b_{ik}}{\lambda_i}\right\|.$$

With the h_k's given by (40) above, we get

$$\|e^{-h_1}, e^{-h_2}, e^{-h_3}, e^{-h_4}, e^{-h_5}\| = \|.1572, .1965, .262, .1965, .1572\|.$$

The entries $F'_k(h_k)b_{3k}$ are given by

$$\|F'_h(h_k)b_{3k}\| = \|.472, .786, 1.31, .786, .472\|.$$

Hence we start with $\lambda_3 = 1.31 \ (= 5\lambda_1/3)$ and get

$$\left\|\frac{F'_k(h_k)b_{ik}}{\lambda_i}\right\| = \left\|\begin{matrix} .2 & .5 & 1 & 1 & 1 \\ 1 & 1 & 1 & .5 & .2 \\ .360 & .6 & 1 & .6 & .360 \end{matrix}\right\|. \tag{41}$$

Since from (39) $\xi_{14} + \xi_{15} < 1$, there is an admissible link from the bottom to the top row and the latter is eligible. Similarly, the second row is eligible, and we therefore begin by lowering λ_1, λ_2, and λ_3 in the constant proportion $.786 : .786 : 1.31 = 3 : 3 : 5$.

During the lowering the matrix (41) remains unchanged, by the same argument as before at formula (32). Hence one of two things can happen. Either $\xi_{14} + \xi_{15} = \xi_{21} + \xi_{22}$ reaches unity, or the sum in the bottom row reaches unity. For the first possibility to occur we must have

$$\frac{1}{5}\log\frac{5}{\lambda_2} + \frac{1}{4}\log\frac{4}{\lambda_2} = 1. \tag{42}$$

7*

But at that point $\xi_{31} = \xi_{32} = \xi_{34} = \xi_{35} = 0$ and

$$\xi_{33} = \frac{h_3}{5} = \frac{1}{5} \log \frac{3}{\lambda_2} < 1$$

from (42). Hence the point (42) is reached first, at, as one easily calculates, $\lambda_1 = \lambda_2 = .479$ and $\lambda_3 = .798$. The matrix (41) now has no admissible links. The solution matrix is now

$$\left\| \begin{matrix} 0 & 0 & 0 & .531 & .469 \\ .469 & .531 & 0 & 0 & 0 \\ 0 & 0 & .367 & 0 & 0 \end{matrix} \right\| . \qquad (43)$$

We have

$$\|h_1, h_2, h_3, h_4, h_5,\| = \|2.346, 2.123, 1.835, 2.123, 2.346\|. \qquad (44)$$

Hence the lowering proceeds with λ_3 alone, and with ξ_{33} the only positive entry in the third row or column. It is given by the formula

$$\xi_{33} = \frac{1}{5} \log \frac{5}{\lambda_3}. \qquad (45)$$

During the lowering the top two 1's in the middle column of the matrix (41) are replaced by smaller numbers. The other elements of the top two rows are unchanged. The sizes of the entries in the 1st, 2nd, 4th and 5th columns in the bottom row are increasing. The entries in the 2nd and 4th place of the bottom row will become unity when

$$\frac{4e^{-h_2}}{\lambda_3} = 1 \qquad (46)$$

the h_2 being given by (44). The solution to (46) is $\lambda_3 = .479$, at which time (45) gives $\xi_{33} = .469$. The solution matrix is now

$$\left\| \begin{matrix} 0 & 0 & 0 & .531 & .469 \\ .469 & .531 & 0 & 0 & 0 \\ 0 & 0 & .469 & 0 & 0 \end{matrix} \right\| .$$

We have

$$\|h_1, h_2, h_3, h_4, h_5\| = \|2.346, 2.123, 2.346, 2.123, 2.346\|.$$

Thus

$$\left\| \frac{F'_k(h_k)b_{ik}}{\lambda_i} \right\| = \left\| \begin{matrix} .2 & .5 & .6 & 1 & 1 \\ 1 & 1 & .6 & .5 & .2 \\ .6 & 1 & 1 & 1 & .6 \end{matrix} \right\| . \qquad (47)$$

There is an admissible link in the fourth column to the top row, and in the second column to the second row. Hence all rows are eligible, and we may lower with $\lambda_1 = \lambda_2 = \lambda_3 = \lambda$. As before, the matrix (47) remains unaltered, and there are only two possible stops. Either $\xi_{21} = \xi_{15}$ reaches unity or $\xi_{52} + \xi_{53} + \xi_{54}$ does.

We shall use formulas (26), (27) and (28) to compute the excesses e_1 and e_2. Denote by λ^0 the common value .479 of λ_1, λ_2 and λ_3 at the outset of the lowering. We get from (26) and (27), applied to the fourth column,

$$\xi^*_{14}(t) = \xi_{14} + \frac{1}{8}\left[\log\frac{4}{\lambda} - \log\frac{4}{\lambda^0}\right]. \tag{48}$$

Applied to the fifth column they give

$$\xi^*_{15}(t) = \xi_{15} + \frac{1}{5}\left[\log\frac{5}{\lambda} - \log\frac{5}{\lambda^0}\right]. \tag{49}$$

Thus

$$e_1(t) = \xi_{14} + \xi_{15} + \frac{1}{8}\left[\log\frac{4}{\lambda} - \log\frac{4}{\lambda^0}\right] + \frac{1}{5}\log\left[\frac{5}{\lambda} - \log\frac{5}{\lambda^0}\right] - 1$$

$$= \frac{1}{8}\left[\log\frac{4}{\lambda} - \log\frac{4}{\lambda^0}\right] + \frac{1}{5}\left[\log\frac{5}{\lambda} - \log\frac{5}{\lambda^0}\right].$$

The formula for $e_2(t)$ is of course identical. Now using (26) and (27)

$$\xi^*_{52}(t) = \xi^*_{54}(t) = \frac{1}{8}\left[\log\frac{4}{\lambda} - \log\frac{4}{\lambda^0}\right]. \tag{50}$$

Now the "exchange rates" λ_1/λ_3 and λ_2/λ_3 are unity. Hence, applying (31), the sum in the third row after transferring the excess is

$$\xi^*_{52}(t) + \xi^*_{53}(t) + \xi^*_{54}(t) + e_1(t) + e_2(t)$$

$$= \frac{1}{2}\left[\log\frac{4}{\lambda} - \log\frac{\lambda}{\lambda^0}\right] + \frac{3}{5}\log\frac{5}{\lambda} - \frac{2}{5}\log\frac{5}{\lambda^0} \tag{51}$$

$$= \frac{1}{2}\log\frac{4}{\lambda} + \frac{3}{5}\log\frac{5}{\lambda} - 2.$$

The value of $\lambda = \lambda_1 = \lambda_2 = \lambda_3$ at which the sum (51) equals unity is .296. At this point $\xi_{53} = \xi_{15} = \xi_{21} = \frac{1}{5}\log\frac{5}{\lambda} = .566$. Applying formula (48) we get $\xi^*_{14} = .591$. The excess in the top row is thus $.591 + .566 - 1 = .157$. This is taken away from the admissible transfer element (1,4) and applied to ξ^*_{54}, which started out, from formula (50), at .060. The result is .217. We thus obtain the solution matrix

$$\|\xi_{ik}\| = \begin{Vmatrix} 0 & 0 & 0 & .434 & .566 \\ .566 & .434 & 0 & 0 & 0 \\ 0 & .217 & .566 & 217 & 0 \end{Vmatrix}. \tag{52}$$

The terminal $\dfrac{F'_k(h_k)b_{ik}}{\lambda_i}$ matrix is given by (47). The values of the h_k are

$$\|h_1, h_2, h_3, h_4, h_5, \| = \|2.83^-, \ 2.60^+, \ 2.83^-, \ 2.60^+, \ 2.83^-\|.$$

We can evidently proceed to cases with more rows. Let us show three possibilities as to the initial linking structure with four weapons. Evidently we may make up an example of $b_{41}, ..., b_{45}$ which, after normalizing with λ_4, gives as the bottom row of the $\dfrac{F'_k(h_k)b_{ik}}{\lambda_i}$ matrix a set $1, x, y, z, w$, where $x, y, z, w < 1$. We would then have

$$\left\| \frac{F'_k(h_k)b_{ik}}{\lambda_i} \right\| = \begin{Vmatrix} .2 & .5 & .6 & 1 & 1 \\ 1 & 1 & .6 & .5 & .2 \\ .6 & 1 & 1 & 1 & .6 \\ 1 & x & y & z & w \end{Vmatrix}. \tag{53}$$

The three admissible links are indicated; admissibility may be checked using (52). Thus all rows are eligible, the admissible transfer element $(1, 4)$ requiring three links to be chained to the bottom row. The other admissible transfer elements are $(2, 1)$ and $(3, 2)$.

If

$$\left\| \frac{F'_k(h_k)b_{ik}}{\lambda_i} \right\| = \begin{Vmatrix} .2 & .5 & .6 & 1 & 1 \\ 1 & 1 & .6 & .5 & .2 \\ .6 & 1 & 1 & 1 & .6 \\ x & 1 & y & z & w \end{Vmatrix} \tag{54}$$

the indicated links are admissible, the rows are all eligible, and the admissible transfer elements are $(1, 4)$, $(2, 2)$, and $(3, 2)$.

If

$$\left\| \frac{F'_k(h_k)b_{ik}}{\lambda_i} \right\| = \begin{Vmatrix} .2 & .5 & .6 & 1 & 1 \\ 1 & 1 & .6 & .5 & .2 \\ .6 & 1 & 1 & 1 & .6 \\ x & y & 1 & z & w \end{Vmatrix} \tag{55}$$

the rows are again all eligible and the admissible transfer elements are $(1, 4)$, $(2, 2)$, and $(3, 3)$.

7. The reversal phenomenon for the marginal utility. Consider the problem

$$\|b_{ik}\| = \begin{Vmatrix} 2a & a \\ b & 2b \end{Vmatrix}$$

with $F_k(h_k) = 1 - e^{-h_k}$. Suppose

$$\frac{1}{2} < b < a \le b + \frac{1}{2} \log 2. \tag{56}$$

There are nine possibilities for the signs of the ξ_{ik}:

(i) $\begin{pmatrix} + & + \\ + & + \end{pmatrix}$ (ii) $\begin{pmatrix} + & + \\ 0 & + \end{pmatrix}$ (iii) $\begin{pmatrix} + & + \\ + & 0 \end{pmatrix}$

(iv) $\begin{pmatrix} + & 0 \\ + & + \end{pmatrix}$ (v) $\begin{pmatrix} + & 0 \\ 0 & + \end{pmatrix}$ (vi) $\begin{pmatrix} + & 0 \\ + & 0 \end{pmatrix}$

(vii) $\begin{pmatrix} 0 & + \\ + & + \end{pmatrix}$ (viii) $\begin{pmatrix} 0 & + \\ 0 & + \end{pmatrix}$ (ix) $\begin{pmatrix} 0 & + \\ + & 0 \end{pmatrix}$.

We first deal with cases (i), (iii), (vii) and (ix). In these cases

$$2ae^{-h_1} \leqq \lambda_1 \qquad ae^{-h_2} = \lambda_1$$
$$be^{-h_1} = \lambda_2 \qquad 2be^{-h_2} \leqq \lambda_2,$$

so that

$$\frac{2a}{b} \leqq \frac{\lambda_1}{\lambda_2} \leqq \frac{a}{2b},$$

from which $4 \leqq 1$, a contradiction.

Now we turn to cases (ii) and (viii). For these

$$2ae^{-h_1} \leqq \lambda_1; \quad ae^{-h_2} = \lambda_1,$$

so that $2e^{-h_1} \leqq e^{-h_2}$ or $h_1 \geqq h_2 + \log 2$. Now

$$h_1 = 2a\xi_{11}; \quad h_2 = a\xi_{12} + 2b.$$

Hence

$$\log 2 + a(1 - \xi_{11}) + 2b \leqq 2a\xi_{11},$$

so that

$$\log 2 + a + 2b \leqq 3a\xi_{11} < 3a.$$

Hence

$$\frac{1}{2}\log 2 + b < a,$$

which contradicts (56).

In cases (iv) and (vi) we have

$$be^{-h_1} = \lambda_2; \quad 2be^{-h_2} \leqq \lambda_2.$$

Hence

$$e^{-h_1} \geqq 2e^{-h_2},$$

so that

$$\log 2 + 2a + b \leqq 3b\xi_{22} \leqq 3b.$$

Hence $a < b$, a contradiction.

The only possible case is therefore (v), so that $h_1 = 2a$ and $h_2 = 2b$.
Hence
$$\lambda_1 = 2ae^{-2a} < 2be^{-2b} = \lambda_2,$$

since ue^{-u} is decreasing for $u > 1$. So $\lambda_1 < \lambda_2$, though $a > b$.

Thus the Lagrange multiplier of what might be throught to be the better weapon is smaller than that of the "inferior" one. Thus in the case at hand it might be misleading as a measure of value. The reversal phenomenon might be thought merely to be an effect of saturation; the "better" weapon has already done all it can do. But the easily demonstrated fact, that in the general problem, whenever $b_{ik} \geq b_{jk}$ for *all* k then $\lambda_i \geq \lambda_j$, shows that this is not the sole reason. The reason for the reversal phenomenon, which in the case at hand says that given a small additional increment of money it is best applied to purchasing the "inferior" weapon, lies in the fact that the "inferior" weapon can deal with some targets which the "better" one cannot — i.e., $2b > a$ in the second row — this coupled with the fact that the "better" has saturated its target.

8. The special case $b_{ik} = a_i c_k$. This case may be treated, from the beginning, in a few lines. The necessary conditions (11) are in the form

$$F_k'(h_k)a_i c_k = \lambda_i \quad \text{if} \quad \xi_{ik} > 0;$$
$$\leq \lambda_i \quad \text{if} \quad \xi_{ik} = 0. \tag{57}$$

Since $\Sigma_k \xi_{ik} = 1$, there is a pair (i, k) for which equality holds in (57). At the same time, the $\leq$ holds for (j, k) with k the same. Thus for some k

$$F_k'(h_k)a_i c_k = \lambda_i$$
$$F_k'(h_k)a_j c_k \leq \lambda_j.$$

On dividing we obtain $\dfrac{a_i}{a_j} \geq \dfrac{\lambda_i}{\lambda_j}$. But evidently some (j, k') has $\xi_{j,k'} > 0$: accordingly the reverse inequality holds. Thus for some γ

$$\frac{a_i}{\lambda_i} = \frac{1}{\gamma} \tag{58}$$

for all i. We may now substitute (58) in (57); we get

$$F_k'(h_k)c_k = \gamma \quad \text{if} \quad \xi_{ik} > 0;$$
$$\leq \gamma \quad \text{if} \quad \xi_{ik} = 0. \tag{59}$$

If $h_k > 0$ some $\xi_{ik} > 0$ and equality holds in (59). If $h_k = 0$ all the $\xi_{ik} = 0$. Hence (59) may be replaced by

$$F_k'(h_k)c_k = \gamma \quad \text{if} \quad h_k > 0;$$
$$\leq \gamma \quad \text{if} \quad h_k = 0. \tag{60}$$

But the solution of (60) is trivial. If $F_k'(0)c_k \leq \gamma$, $h_k = 0$. If $F_k'(0)c_k > \gamma$, the value of h_k is the solution of $F_k'(h_k)c_k = \gamma$.

Thus the values of the h_k have been found, as soon as we know the value of γ. From the formula

$$h_k = \Sigma_i a_i c_k \xi_{ik}$$

we get

$$\frac{h_k}{c_k} = \Sigma_i a_i \xi_{ik},$$

so that

$$\Sigma_k \frac{h_k}{c_k} = \Sigma_k \Sigma_i a_i \xi_{ik} = \Sigma_i a_i. \tag{61}$$

Evidently each h_k is a constant or decreasing function of γ. Thus to obtain the value of γ which solves our problem we need only start with $\gamma_0 = \underset{k}{\mathrm{Max}} F'_k(0) \cdot c_k$ and start lowering, until equation (61) is satisfied.

The ξ_{ik} are far from unique, as one could deduce from a study of the variants on the "program" of Section 4 possible in this case. One solution, which may be found by inspection, is

$$\xi_{ik} = \frac{h_k/c_k}{\Sigma_k h_k/c_k}. \tag{62}$$

To verify this one notes that obviously the ξ_{ik} sum to unity on k, and

$$\Sigma_i b_{ik} \xi_{ik} = \Sigma_i a_i c_k \cdot \frac{h_k}{c_k} \cdot \frac{1}{\Sigma_k (h_k/c_k)}$$

$$= \frac{h_k}{\Sigma_k (h_k/c_k)} \cdot \Sigma_i a_i = h_k,$$

the last equality holding from (61).

Thus the "separable" problem $b_{ik} = a_i c_k$ is easily solved by hand; it is as easy as the one-weapon problem. We shall nevertheless see in Chapter VII that it fits a quite important realistic case as well as can be done without going to integers.

Exercises to Chapter VI

1. Prove that conditions (11) are sufficient to ensure that the matrix ξ is a solution. An elementary proof can be made along the lines of the proof of Lemma 9 in Chapter II.

2. Write out the proof of the detail in Lemma 4, that if an element of the jth row, $j \neq i$, is linked to anything it is in Z_j.

3. Verify the statements in case (ii) at the end of Section 4. Use Lemma 5.

4. (Difficult). What happens to λ_i when a row, say the nth, is multiplied through by a constant near unity? In particular, replace b_{nk} by $r b_{nk}$, $k = 1, \ldots, K$, where $r > 1$ and r is close to 1. Now replace λ_n by $r \lambda_n$ and ξ_{nk} by $\frac{1}{r} \xi_{nk}$. Then the matrix

$$\left\| \frac{F'_k(h_k) b_{ik}}{\lambda_i} \right\|$$

still satisfies condition (11), except that $\Sigma \xi_{nk} = \frac{1}{r}$. One may now lower to overcome this deficit. Holding r fixed, one calculates $\dot{h}_k(t)$,

$\varrho_k(t)$, $\dot{e}_i(t)$, (including $\dot{e}_n(t)$; this is the "local rate of accumulation"), and finally $(\lambda_i/r\lambda_n)\,\dot{e}_i(t)$, and thus obtains the total rate at which accumulation is taking place in the nth row. In this way one can find the "time" required to make up the deficit. The rate at which accumulation is taking place is

$$R(t) = \frac{1}{t^2 r \lambda_n} \sum_{k \in S} \frac{[F_k'(h_k(t))]^2}{-F_k''(h_k(t))}, \tag{63}$$

where S denotes the set of k for which h_k is varying during the lowering. Suppose it takes until $t(r)$ to make up the deficit. We must have, by the law of the mean, for some t' with $t(r) < t' < 1$

$$R(t')\,[1 - t(r)] = 1 - \frac{1}{r},$$

so that

$$\frac{t(r) - 1}{r - 1} = \frac{-1}{rR(t')}.$$

We thus find that the right derivative of t with respect to r at $r = 1$ is

$$D_R t(1) = \frac{-1}{R(1)}.$$

Since $\lambda_n(t) = rt\lambda_n$ and $\lambda_i(t) = t\lambda_i$, $i \neq n$, we get the following formulas for the right derivative $D_R \lambda_i$ and $D_R \lambda_n$ with respect to r: if the ith row is eligible,

$$\frac{1}{\lambda_i} D_R \lambda_i = \frac{-\lambda_n}{\displaystyle\sum_{k \in S} \frac{F_k'^2}{-F_k''}};$$

and

$$\frac{1}{\lambda_n} D_R \lambda_n = 1 - \frac{\lambda_n}{\displaystyle\sum_{k \in S} \frac{F_k'^2}{-F_k''}}.$$

How does the set S change when taking the left derivative at $r = 1$?

5. *Finiteness of the process.* For the lowering process of section 4 one may derive a formula analogous to (63) (with $r = 1$) for the rate of accumulation in the nth row. One may thus calculate the time $t_{(iii)}$ at which stop (iii) would take place. Similarly one may calculate times $t_{(i)}$ and $t_{(ii)}$. The point to which one lowers is then the maximum t_* of those times. t_* is definitely less than unity.

Now suppose that we have reached a stage at which $\Sigma_k \xi_{nk} = w$, with $0 < w < 1$. Then one may consider the problem of *lowering* w, i.e. of *raising* λ_n. For this one has an obvious new definition of the admissible linking structure. One then has a first stop t^* with t^* definitely larger than unity. Prove from these considerations, or otherwise, that the lowering process will take only finitely many steps.

Chapter VII

On stability and Max-Min-Max

We have described in Chapter I a problem in which one side buys a mixture of weapons systems, his opponent discovers what he has done and buys a countering mixture, and finally the original side allocates his residuum to the opponent's target complex. There are three stages: Max-Min-Max, in which the original side has the two outside Max moves and his opponent the middle Min move.

The general Max-Min-Max problem leads directly (see Section 6) to the concepts of strong and weak forward and backward stability. These concepts are therefore defined and discussed first in Section 1. They apply to Max-Min problems more generally and are illustrated on the examples of Chapter IV. The question of stability is essential to any Max-Min problem.

The Max-Min-Max problem of particular interest to us is presented in Section 2. Some differentiations, which in the percentage vulnerable case reduce it to a Max-Min problem, are carried out in Section 3. In Section 4 necessary conditions for a solution are stated. In Section 5 there is a sufficiency criterion for a system to be admitted to a mix, for the general percentage vulnerable case. Section 6 gives a theorem which shows the remarkable consequences of weak or strong forward stability in the case of percentage vulnerable systems: if strong forward stability holds, the Max-Min-Max problem is completely solved, at least as to the most important, "outside," problem. Section 7 gives a Max-Min-Max example which is strongly backward stable and not weakly forward stable; the conclusion of Theorem III does not hold in this case.

Section 8, corresponding to Section 8 of Chapter VI, completely solves the Max-Min-Max problem for the case of separable coefficients. In this case the calculations are easily done by hand. This is by far the easiest section of this chapter. Nevertheless the separable case appears useful in current practice; see the concluding remarks concerning point targets.

1. *Strong and weak forward and backward stability.* These concepts are defined only for admissible directions γ, i.e. with $\Sigma \gamma_i = 0$. Let I^+ denote the set of i for which $\gamma_i > 0$, and I^- the set for which $\gamma_i < 0$.

By *strong forward stability* in the direction γ at the point x we mean the following. There exist an $i \in I^+$ and sequences $\{x^m\}$ and $\{y^m\}$ with $x^m \to x$ from the direction γ, $y^m \in Y(x^m)$, $y^m \to y^\gamma$ for some $y^\gamma \in Y^\gamma(x)$, and $y_i^m > 0$ for all m.

Weak forward stability means that given that $y_{i'} > 0$ for some $i' \in I^+$ and $y \in Y(x)$, then there are an $i \in I^+$ and sequences $\{x^m\}$ and $\{y^m\}$ as in the definition of strong forward stability.

By *strong backward stability* in the direction γ at the point x we mean the following. There exist an $i \in I^-$ and sequences $\{x^m\}$ and $\{y^m\}$ with $x^m \to x$ from the direction γ and $y^m \in Y(x^m)$, $y^m \to y^\gamma$ for some $y^\gamma \in Y^\gamma(x)$, with $y_i^m > 0$ for all m.

Weak backward stability means that given $y_{i'} > 0$ for some $i' \in I^-$ and $y \in Y(x)$, then there are an $i \in I^-$ and sequences $\{x^m\}$ and $\{y^m\}$ as in the definition of strong backward stability.

We turn to the examples of Chapter IV. In the example of Section 1, at the solution point $x_1 = x_2 = 1/2$ backward stability (the concepts of strong and weak backward stability coincide here) holds for both admissible γ (Exercise 1).

In the allocation game treated in Chapter II [formula (9)] and in Section 2 of Chapter IV, there is weak forward and backward stability at any point $x = (x_1, \ldots, x_n)$ and in any admissible direction (Exercise 2).

In the problem of Section 3 of Chapter IV there is weak forward and backward stability if $\beta_i < 1$. In the $n = 2$ numerical example there, in which $\beta_i > 1$, there was strong backward stability (Exercise 3).

Finally, in the example of Section 4 of Chapter IV, most of the work of solution involved establishing strong forward stability; this was completed in Lemma 4, and was the key to the solution.

We turn to the formulation of the general Max-Min-Max problem.

2. *A Max-Min-Max problem.* We suppose that there are three moves. The maximizing player chooses a strategy $x = (x_1, \ldots, x_n)$. Thereupon and in full knowledge of x the minimizing player chooses a strategy $y = (y_1, \ldots, y_m)$ which acts on x according to formulas of percentage vulnerable, numerically vulnerable, or simultaneously (percentage) vulnerable type, as in Chapter V, to yields a vector $x' = (x_1', \ldots, x_n')$. In all these cases, the x_i' are continuous and differentiable functions of x and y. The maximizing player now chooses a matrix $\xi = \|\xi_{ik}\|$, $i = 1, \ldots, n$; $k = 1, \ldots, K$, with

$$\sum_{k=1}^{K} \xi_{ik} = 1, \quad i = 1, \ldots, n \tag{1}$$

and all $\xi_{ik} \geq 0$. The entry ξ_{ik} is the proportion of his residuum of the ith weapon which he sends to the kth target. Thus the total amount of the ith weapon arriving at that target is $x_i' \xi_{ik}$. This time we shall denote by β_{ik} the effectiveness coefficient of one money unit's worth of the ith weapon against the kth target. The "total weight of attack," as in Chapter VI, on the kth target is now

$$h_k = \sum_{i=1}^{n} \beta_{ik} x_i' \xi_{ik}. \tag{2}$$

For each set x_i' we write

$$b_{ik} = x_i' \beta_{ik} \tag{3}$$

and thus identify (2) with the h_k of Chapter VI, where b_{ik} was the effectiveness of one *proportional* unit of the total amount of the ith weapon. Let us write

$$D(x, y, \xi) = \Sigma F_k(h_k) \tag{4}$$

with the general F_k of Chapter VI. The problem is to calculate

$$\operatorname*{Max}_{x} \operatorname*{Min}_{y} \operatorname*{Max}_{\xi} D(x, y, \xi). \tag{5}$$

We now turn to the percentage vulnerable case.

3. *Some differentiations (Percentage vulnerable case).* We begin by studying the problem of differentiating

$$H(x, y) = \operatorname*{Max}_{\xi} D(x, y, \xi) \tag{6}$$

with respect to x and y in the percentage vulnerable case. Recall the necessary condition (11) of Chapter VI:

$$\begin{aligned} F_k'(h_k)\beta_{ik}x_i e^{-\alpha_i y_i} &= \lambda_i \quad \text{if} \quad \xi_{ik} > 0; \\ &\leq \lambda_i \quad \text{if} \quad \xi_{ik} = 0. \end{aligned} \tag{7}$$

Recall that the λ_i and h_k are uniquely determined by the coefficient matrix $\|b_{ik}\| = \|\beta_{ik}x_i'\|$. From this fact and the continuity of the functions $F_k'(h_k)$ it follows (Exercise 4) that the $\lambda_i = \lambda_i(x, y)$ and $h_k = h_k(x, y)$ appearing in (7) are continuous functions of x and y.

When $x_i > 0$ we have from (7):

$$\frac{\lambda_i}{x_i} = \operatorname*{Max}_{k} F_k'(h_k)\beta_{ik}e^{-\alpha_i y_i}. \tag{8}$$

For notational convenience we shall write λ_i/x_i to mean the right hand side of (8) also in the case $x_i = 0$.

Now $D(x, y, \xi)$ is continuously differentiable in the x_i and y_i, the continuity being in the x, y, ξ space. Hence we may apply Theorem I of Chapter III to compute the directional derivatives of $H(x, y)$ in directions in the x- and y- spaces. We get, if γ is a direction in the x-space,

$$D_\gamma H(x, y) = \operatorname*{Max}_{\xi \in \Xi} \sum_{i=1}^{n} \gamma_i \frac{\partial}{\partial x_i} [D(x, y, \xi)], \tag{9}$$

where Ξ denotes the "answering set" of ζ's yielding the maximum for fixed x and y. Hence

$$D_y H(x, y) = \underset{\xi \in \Xi}{\text{Max}} \sum_{i=1}^{n} \sum_{k=1}^{K} \gamma_i F_k'(h_k) e^{-\alpha_i y_i} \beta_{ik} \xi_{ik}$$

$$= \underset{\xi \in \Xi}{\text{Max}} \sum_{i=1}^{n} \sum_{k=1}^{K} \frac{\gamma_i \lambda_i}{x_i} \xi_{ik} \tag{10}$$

$$= \underset{\xi \in \Xi}{\text{Max}} \sum_{i=1}^{n} \frac{\gamma_i \lambda_i}{x_i},$$

where the passage from the top line to the second line took account of the convention (8). The reader should verify that this step is legitimate (Exercise 5). But we have seen that, though $\xi \in \Xi$ is not unique, the $\lambda_i = \lambda_i(x, y)$ are unique. Hence (10) yields

$$D_y H(x, y) = \sum_{i=1}^{n} \frac{\gamma_i \lambda_i(x, y)}{x_i}. \tag{11}$$

But there is more. The quantities $\lambda_i(x, y)$ are also continuous. In fact, so are the quantities $\dfrac{\lambda_i(x, y)}{x_i}$, when $x_i = 0$ as well (Exercise 6). It follows that $D_y H(x, y)$ is continuous in the space of pairs (x, y). But in this case (recall the proof of Theorem II, Chapter III) $H(x, y)$ is continuously differentiable in the ordinary sense and the elementary chain rule holds:

$$D_y H(x, y) = \sum_{i=1}^{n} \gamma_i H_{x_i}(x, y), \tag{12}$$

where $H_{x_i}(x, y)$ denotes the partial derivative with respect to x_i. It is left as an exercise for the reader (Exercise 7) to compute $D_g(x, y)$ in a direction g parallel to the y-space. The result is

$$D_g H(x, y) = - \sum_{i=1}^{n} g_i \alpha_i \lambda_i(x, y), \tag{13}$$

from which one again deduces that $H(x, y)$ is continuously differentiable in the ordinary sense and that

$$D_g H(x, y) = \sum_{i=1}^{n} g_i H_{y_i}(x, y). \tag{14}$$

We have thus proved the following.

Theorem I. $H(x, y)$ *as defined by* (6) *is continuously differentiable in the ordinary sense with respect to all its variables, and*

$$H_{x_i}(x, y) = \frac{\lambda_i(x, y)}{x_i}; \tag{15}$$

$$H_{y_i}(x, y) = \alpha_i \lambda_i(x, y). \tag{16}$$

In formula (15) *we recall the convention* (8).

What is striking about Theorem I is that $H(x, y)$ is defined as $\underset{\xi}{\text{Max}} D(x, y, \xi)$, and the set ξ yielding this maximum is not unique.

We are now ready to proceed. $H(x, y)$ itself satisfies all the requirements of Theorem I, Chapter III. In fact put

$$\varphi(x) = \underset{y}{\text{Min }} H(x, y)$$

where the y is subjected to $\Sigma y_i = 1$, $y_i \geq 0$. Then $D_\gamma \varphi(x)$ exists and is given by

$$D_\gamma \varphi(x) = \underset{y \in Y(x)}{\text{Min}} \sum_{i=1}^{n} \gamma_i H_{x_i}(x, y) \tag{17}$$

$$= \underset{y \in Y(x)}{\text{Min}} \sum_{i=1}^{n} \frac{\gamma_i \lambda_i(x, y)}{x_i}.$$

4. *The necessary conditions in the percentage vulnerable case.*

With formulas (17) and (16) in hand, it is easy to state the necessary conditions for a solution. These are that

$$\underset{y \in Y(x)}{\text{Min}} \Sigma \gamma_i \lambda_i / x_i \leq 0 \tag{18}$$

for any direction γ satisfying $\Sigma \gamma_i = 0$ and $\gamma_i \geq 0$ when $x_i = 0$, and that for the y which yields the minimum on the left hand side of (18), or for that matter for any $y \in Y(x)$, there exists a μ such that

$$\begin{aligned} \lambda_i \alpha_i &= \mu \quad \text{if} \quad y_i > 0; \\ &\leq \mu \quad \text{if} \quad y_i = 0. \end{aligned} \tag{19}$$

5. *A sufficiency criterion for admission to the mix in the percentage vulnerable case.*

We shall give the criterion in the following form. Suppose the Max-Min-Max problem solved for $(n-1)$ systems. Thus there is a set $x_1, ..., x_{n-1}$ with $\sum_{i=1}^{n-1} x_i = 1$ and $x_i \geq 0$, and a collection $y \in Y(x_1, ..., x_{n-1})$ of answering strategies, and a collection $\Xi(x_1, ..., x_{n-1}; y)$ of answering matrices ξ for each y. There is a λ corresponding to the constraint on the x (Theorem VI of Chapter III), and for each answering strategy y an associated μ and uniquely determined sets $\lambda_1, ..., \lambda_{n-1}; h_1, ..., h_K$. It is in terms of some of these parameters that we shall state the condition for x_n to be positive in an optimal mix involving all n systems.

Suppose $x_n = 0$ in an optimal mix involving n systems. Then, for any answering y, $\lambda_n(x, y) = 0$ by formula (7). From (19), $y_n = 0$ as well. Consider the direction γ with $\gamma_n = 1$, all other $\gamma_i = 0$. From the Lagrange multiplier principle for Max-Min (Theorem VI, Chapter III) we find that

$$D_\gamma \varphi(x) \leq \lambda. \tag{20}$$

From formula (17) and the convention (8) we get

$$D_\gamma \varphi(x) = \underset{y \in Y(x)}{\text{Min}} \frac{\lambda_n}{x_n}$$
$$= \underset{y \in Y(x)}{\text{Min}} \left[\underset{k}{\text{Max}} F_k'(h_k) \beta_{nk} e^{-\alpha_n y_n} \right] \qquad (21)$$
$$= \underset{y \in Y(x)}{\text{Min}} \left[\underset{k}{\text{Max}} F_k'(h_k) \beta_{nk} \right],$$

where we recall that the set $\{h_k\}$ depends on y. Hence, from (20), if $x_n = 0$ we have

$$\underset{y \in Y(x)}{\text{Min}} \; \underset{k}{\text{Max}} \; [F_k'(h_k) \beta_{nk}] \leqq \lambda . \qquad (22)$$

Thus a sufficiency criterion may be stated as follows: if for all $y \in Y(x_1, \dots, x_{n-1})$ we have

$$\underset{k}{\text{Max}} F_k'(h_k) \beta_{nk} > \lambda , \qquad (23)$$

then $x_n > 0$ in the optimal mix of n systems.

It is not so easy to go the other way. No corresponding necessity criterion is known to the author (Exercise 8).

6. *Consequences of strong and weak forward stability in the Max-Min-Max problem for the percentage vulnerable case.*

These consequences are remarkable, as we shall see.

Theorem II. *If the solution to the (percentage-vulnerable) problem* $\underset{y}{\text{Min}} \; H(x, y)$ *subject to* $\Sigma y_i = 1$, $y_i \geqq 0$ *is strongly forward stable in any admissible direction at the point* x^0 *yielding* $\text{Max Min} H(x, y)$ *subject to* $\Sigma x_i = 1$, $x_i \geqq 0$, *then there is a constant* k *such that*

$$x_i^0 = \frac{k}{\alpha_i} \qquad (24)$$

whenever $x_i^0 > 0$.

Proof. Suppose $x_i^0 > 0$ and $x_j^0 > 0$, $j \neq i$. Consider the direction γ defined by $\gamma_i = -1/\sqrt{2}$, $\gamma_j = +1/\sqrt{2}$, $\gamma_s = 0$ for $s \neq i, j$. Then there exists by strong forward stability a $y^\gamma \in Y^\gamma(x)$ with the following property. There is a sequence $x^m \to x^0$ with $x_i^m < x_i^0$, $x_j^m > x_j^0$, and $x_s^m = x_s^0$, $s \neq i, j$, and a sequence $y^m \in Y(x^m)$ with $y_j^m > 0$ for all j and with $y^m \to y^\gamma \in Y^\gamma(x)$. From Theorem I of Chapter III,

$$\sqrt{2} D_\gamma \varphi(x^0) = \frac{\lambda_j(x^0, y^\gamma)}{x_j^0} - \frac{\lambda_i(x^0, y^\gamma)}{x_i^0} . \qquad (25)$$

Since $D_\gamma \varphi(x^0) \leqq 0$ because γ is an admissible direction, we have

$$\frac{\lambda_j(x^0, y^\gamma)}{x_j^0} \leqq \frac{\lambda_i(x^0, y^\gamma)}{x_i^0} . \qquad (26)$$

Now since $y_j^m > 0$ for each m, we have

$$\alpha_j \lambda_j(x^m, y^m) \geqq \alpha_i \lambda_i(x^m, y^m).$$

On taking the limit as $m \to \infty$, we get

$$\alpha_j \lambda_j(x^0, y^\gamma) \geqq \alpha_i \lambda_i(x^0, y^\gamma). \tag{27}$$

On dividing (27) by (26) we get

$$x_j^0 \alpha_j \geqq x_i^0 \alpha_i. \tag{28}$$

Since $x_j^0 > 0$ we may now take the direction with $\gamma_i = 1/\sqrt{2}$, $\gamma_j = -1/\sqrt{2}$ and obtain the inequality

$$x_i^0 \alpha_i \geqq x_j^0 \alpha_j \tag{29}$$

(28) combined with (29) yields the theorem.

Theorem III. *If strong forward stability in the hypothesis of Theorem II is replaced by weak forward stability, then there is a constant k such that*

$$x_i^0 = \frac{k}{\alpha_i}$$

whenever $y_i > 0$ for some $y \in Y(x^0)$.

The proof of Theorem III is left to the reader (Exercise 9).

Thus strong and weak forward stability are seen to have very striking consequences. If strong forward stability holds the percentage-vulnerable Max-Min-Max problem is completely solved by Theorem II. It says that once a system is admitted to the mix (for which a sufficient condition was given in Section 5) it is admitted in a quantity inversely proportional to its vulnerability and not any more depending on its effectiveness. Theorem III says that once a system is admitted to the mix and is, so to speak, worth attacking, then it is admitted in the same amount. The game of formula (9) of Chapter II shows that the formula $x_i^0 = k/\alpha_i$ does not necessarily hold when all the $y_i = 0$. This is the case $v_i = \lambda$; recall Lemma 9 of Chapter II.

7. *An example of the failure of weak forward stability.* This section treats an example which shows that weak forward stability is not a general property of Max-Min-Max problems. The example in fact is strongly backward stable at the solution point.

We take the matrix

$$\|\beta_{ik}\| = \left\|\begin{matrix} 10 & 1 \\ 1 & 10 \end{matrix}\right\|. \tag{30}$$

Thus, from (3), for percentage vulnerable systems

$$\|b_{ik}\| = \left\|\begin{matrix} 10x_1 e^{-\alpha_1 y_1} & x_1 e^{-\alpha_1 y_1} \\ x_2 e^{-\alpha_2 y_2} & 10x_2 e^{-\alpha_2 y_2} \end{matrix}\right\|. \tag{31}$$

We take $v_1 = v_2 = 1$, $F_k(h_k) = 1 - e^{-h_k}$, $\alpha_1 = 1$, and $\alpha_2 = \alpha$ with α near unity.

8 Danskin, The Theory of Max-Min

We first observe that the inside solution matrix ξ is of one of three types:

$$\begin{pmatrix} + & + \\ 0 & + \end{pmatrix};$$ (32)

$$\begin{pmatrix} + & 0 \\ + & + \end{pmatrix};$$ (33)

$$\begin{pmatrix} + & 0 \\ 0 & + \end{pmatrix}.$$ (34)

This follows from the condition (11) of Chapter VI, which here may be written in the form

$$
\begin{aligned}
10x_1 e^{-y_1} e^{-h_1} &= \lambda_1 \quad \text{if} \quad \xi_{11} > 0, \\
&\leq \lambda_1 \quad \text{if} \quad \xi_{11} = 0; \quad \text{(a)} \\
x_1 e^{-y_1} e^{-h_2} &= \lambda_1 \quad \text{if} \quad \xi_{12} > 0, \\
&\leq \lambda_1 \quad \text{if} \quad \xi_{12} = 0; \quad \text{(b)} \\
x_2 e^{-\alpha y_2} e^{-h_1} &= \lambda_2 \quad \text{if} \quad \xi_{21} > 0, \\
&\leq \lambda_2 \quad \text{if} \quad \xi_{21} = 0; \quad \text{(c)} \\
10x_2 e^{-\alpha y_2} e^{-h_2} &= \lambda_2 \quad \text{if} \quad \xi_{22} > 0, \\
&\leq \lambda_2 \quad \text{if} \quad \xi_{22} = 0. \quad \text{(d)}
\end{aligned}
$$ (35)

Suppose that $x_2 > 0$ and both $\xi_{21} > 0$ and $\xi_{12} > 0$. Then from (35a) and (35c) we get

$$\frac{10x_1 e^{-y_1}}{x_2 e^{-\alpha y_2}} \leq \frac{\lambda_1}{\lambda_2},$$ (36)

and from (35b) and (35d) we get

$$\frac{x_1 e^{-y_1}}{10x_2 e^{-\alpha y_2}} \geq \frac{\lambda_1}{\lambda_2}.$$ (37)

Evidently (37) and (36) are in contradiction. The cases $\begin{pmatrix} + & 0 \\ + & 0 \end{pmatrix}$ and $\begin{pmatrix} 0 & + \\ 0 & + \end{pmatrix}$ are ruled out similarly.

The case $x_2 = 0$ is trivial: in this case the values of ξ_{21} and ξ_{22} are immaterial, and one solution must be of the form (32), (33), or (34).

We shall proceed as follows. For each value of x_1, x_2 we shall seek the pairs y_1, y_2 yielding $\underset{y}{\text{Min}} \underset{\xi}{\text{Max}} D(x, y, \xi)$. Now the pair y_1, y_2 must lead to a solution ξ of one of the three types (32), (33), (34). We are therefore led first to ask for the y-solution to $\underset{y}{\text{Min}} \underset{\xi}{\text{Max}} D(x, y, \xi)$ under the restriction that the resulting solution for ξ should be of the form (32). The next four paragraphs are devoted to this case.

If then the solution is of the form (32), we must have $10e^{-h_1} = e^{-h_2}$, so that

$$h_1 = h_2 + \log 10. \tag{38}$$

Now

$$h_1 = \xi_{11} 10 x_1 e^{-y_1} \tag{39}$$

and

$$h_2 = \xi_{12} x_1 e^{-y_1} + 10 x_2 e^{-\alpha y_2}. \tag{40}$$

We then get

$$\xi_{11} = \frac{x_1 e^{-y_1} + 10 x_2 e^{-\alpha y_2} + \log 10}{11 x_1 e^{-y_1}}, \tag{41}$$

$$\xi_{12} = \frac{10 x_1 e^{-y_1} - 10 x_2 e^{-\alpha y_2} - \log 10}{11 x_1 e^{-y_1}}, \tag{42}$$

$$h_1 = \frac{10 x_1 e^{-y_1} + 100 x_2 e^{-\alpha y_2} + 10 \log 10}{11}, \tag{43}$$

$$h_2 = \frac{10 x_1 e^{-y_1} + 100 x_2 e^{-\alpha y_2} - \log 10}{11}. \tag{44}$$

Since $h_1 = h_2 + \log 10$, the problem for y of minimizing

$$D(x, y, \xi) = 1 - e^{-h_1} + 1 - e^{-h_2}$$

with x_1, x_2 fixed and ξ_{11}, ξ_{22} given by (41) and (42), is the same as minimizing h_1 as given by (43), i.e. the same as that of minimizing

$$x_1 e^{-y_1} + 10 x_2 e^{-\alpha y_2}. \tag{45}$$

At the same time, we must observe the restriction that $\xi_{12} > 0$ in (42), i.e.

$$x_1 e^{-y_1} > x_2 e^{-\alpha y_2} + \frac{\log 10}{10}. \tag{46}$$

Now (46) implies that

$$x_1 > x_2 e^{-\alpha} + \frac{\log 10}{10}. \tag{47}$$

i.e. that

$$x_1 > \frac{10 e^{-\alpha} + \log 10}{10(1 + e^{-\alpha})} = x_1^* \doteq .437 \tag{48}$$

for α near unity. Now suppose that x_1 satisfies

$$x_1^* < x_1 \leqq \frac{10 \alpha e^{-\alpha}}{1 + 10 \alpha e^{-\alpha}} \doteq .786 \tag{49}$$

(α always near unity). Suppose that $y_1 > 0$ in the solution to the minimum problem (45). Then, for some μ, $x_1 e^{-y_1} = \mu$ and $10 \alpha x_2 e^{-\alpha y_2} \leqq \mu$.

8*

These imply

$$x_1 e^{-y_1} \geq 10\alpha x_2 e^{-\alpha y_2},$$

so that

$$x_1 > 10\alpha x_2 e^{-\alpha},$$

which contradicts (49). It follows that on the interval (49) the y-solution to (45) has $y_1 = 0$, so that the value of the minimum is

$$x_1 + 10 x_2 e^{-\alpha}. \tag{50}$$

The quantity (50) is evidently decreasing for x_1 on the interval (49). Also, since $y_1 = 0$, the inequality (46) is *equivalent* to the inequality (47), so that indeed $\xi_{12} > 0$ for x_1 on (49).

Now turn to the interval

$$\frac{10\alpha e^{-\alpha}}{1 + 10\alpha e^{-\alpha}} < x_1 < \frac{10\alpha}{10\alpha + e^{-1}} \doteq .964. \tag{51}$$

Then exists a $\mu > 0$ such that

$$\begin{array}{llll} x_1 e^{-y_1} = \mu & \text{if} \quad y_1 > 0, & 10\alpha x_2 e^{-\alpha y_2} = \mu & \text{if} \quad y_2 > 0, \\ \quad\quad \leq \mu & \text{if} \quad y_1 = 0; & \quad\quad\quad\quad\quad \leq \mu & \text{if} \quad y_2 = 0. \end{array} \tag{52}$$

We easily see from (52) that if $y_2 = 0$ we would have $x_1 \geq 10\alpha(10\alpha + e^{-1})$ and that if $y_1 = 0$ we would have $x_1 \leq 10\alpha e^{-\alpha}/(1 + 10\alpha e^{-\alpha})$. Hence on the interval (51) both $y_1 > 0$ and $y_2 > 0$; equality holds in (52), and the value of μ is given by the equation

$$\log \frac{x_1}{\mu} + \frac{1}{\alpha} \log \frac{10\alpha x_2}{\mu} = 1. \tag{53}$$

We find from (53) that the derivative of μ with respect to x_1 is given by

$$\mu' = \frac{\mu \left(\dfrac{1}{x_1} - \dfrac{1}{\alpha x_2} \right)}{\left(1 + \dfrac{1}{\alpha} \right)},$$

so that on the interval (51) μ is strictly decreasing. The quantity (45) is now given by

$$\mu \left(1 + \frac{1}{\alpha} \right);$$

hence (45) is strictly decreasing on the interval (51). We now consider the inequality (46) on the interval (51). This reads

$$\mu > \frac{\mu}{10\alpha} + \frac{\log 10}{10},$$

i.e.

$$\mu\left(1 - \frac{1}{10\alpha}\right) > \frac{\log 10}{10}. \tag{54}$$

At the right end point of (51) we get from (53) that

$$\mu^2 \doteq \frac{10 x_1 x_2}{e} \doteq .343$$

for $\alpha \doteq 1$. Thus $\mu \doteq .586$ there. Since μ is strictly decreasing on (51) the condition (54), i.e. the condition (46), holds on (51) with a margin sufficient to allow small variations of α around unity.

We finally turn to the interval

$$\frac{10\alpha}{10\alpha + e^{-1}} \leqq x_1 \leqq 1. \tag{55}$$

It follows from (52) that $y_2 = 0$ on (55); if y_2 were positive we would have $10\alpha x_2 e^{-\alpha y_2} \geqq x_1 e^{-y_1}$, so that $10\alpha x_2 > x_1 e^{-1}$, which contradicts (55). Hence (45) is given by

$$x_1 e^{-1} + 10 x_2,$$

which is evidently decreasing on (55). The condition (46) reads

$$x_1 e^{-1} > x_2 + \frac{1}{10} \log 10, \tag{56}$$

or

$$x_1 > \frac{10 + \log 10}{10(1 + e^{-1})} \doteq .899. \tag{57}$$

Since the left endpoint of (55) is $10\alpha/(10\alpha + e^{-1}) \doteq .964$, (57), and thus also (46), evidently hold on (55).

We have thus found that on the entire interval $(x_1^*, 1) \doteq (.437, 1)$ there is a unique solution for y corresponding to the possibility (32); that throughout this interval the problem for y of minimizing $D(x, y, \xi)$ with respect to the possibility (32) is equivalent to that of minimizing h_1; and that the quantity h_1 corresponding to the y-solution is strictly decreasing on $(x_1^*, 1)$. Off that interval there is no solution for y which leads to a solution of type (32) for ξ.

We find similarly that on the interval $(0, x_1^{**})$, where

$$x_1^{**} = \frac{10 - \log 10}{10(1 + e^{-1})} \doteq .563, \tag{58}$$

there is a unique solution for y leading to a solution for ξ of type (33), and that the value of the corresponding h_1 is strictly increasing on $(0, x_1^{**})$.

In this case

$$h_2 = h_1 + \log 10,$$

$$\xi_{21} = \frac{10x_2 e^{-\alpha y_2} - 10x_1 e^{-y_1} - \log 10}{11 x_2 e^{-\alpha y_2}}, \tag{59}$$

$$\xi_{22} = \frac{10x_1 e^{-y_1} + x_2 e^{-\alpha y_2} + \log 10}{11 x_2 e^{-\alpha y_2}}, \tag{60}$$

$$h_1 = \frac{10x_2 e^{-y_2} + 100x_1 e^{-l_1} - \log 10}{11}, \tag{61}$$

and

$$h_2 = \frac{10x_2 e^{-y_2} + 100x_1 e^{-y_1} + 10\log 10}{11}. \tag{62}$$

We note that on the interval

$$.214 = \frac{1}{1 + 10e^{-1}} < x_1 < x_1^{**} \tag{63}$$

the solution for y corresponding to the possibility (33) has $y_1 = 1$ and $y_2 = 0$.

The third possibility is that the y corresponds to (34). This situation was worked out completely in Section 3 of Chapter IV for the case $\alpha = \alpha_2 = 1$. We found there that on the interval $(e/10, 1 - e/10)$, the resulting y-problem was strictly concave, and thus the solution either $y = (1, 0)$ or $y = (0, 1)$; the y yielding the minimum on $(e/10, 1/2)$ is $(1, 0)$ and, on $(1/2, 1 - e/10)$, $(0, 1)$. But we have just seen (63) that on the interval

$$.214 \doteq \frac{1}{1 + 10e^{-1}} < x_1 < x_1^{**} = \frac{10 - \log 10}{10(1 + e^{-1})} \doteq .563$$

the choice of $y = (1, 0)$ leads to a maximizing matrix with a unique solution having $\xi_{21} > 0$; accordingly on $(e/10, 1/2)$ the y-solution corresponding to the diagonal matrix leads to a quantity strictly below that yielded by the possibility (33). But since, as we saw in Chapter IV, the maximum corresponding to the diagonal possibility in fact occurs at $x_1 = x_2 = 1/2$, then the diagonal possibility cannot occur at all. The solution is now obviously at the intersection of the curves for $D(x, y, \xi)$ corresponding to the possibilities (32) and (33), which is $x_1 = x_2 = 1/2$.

At the point $(1/2, 1/2)$ we calculate, using the possibility (33) [formula (61)]

$$h_1 = \frac{50e^{-1} + 5 - \log 10}{11} \doteq 1.917;$$

$$e^{-h_1} \doteq .147;$$

$$e^{-h_2} = .1e^{-h_1} \doteq .0147.$$

Thus for $x_1 = x_2 \doteq .5$

$$D(x, y, \xi) = 1 - e^{-h_1} + 1 - e^{-h_2} \doteq 1.838 \,. \tag{64}$$

If the minimizing strategy against $x_1 = x_2 = 1/2$ is in fact $(1, 0)$, the optimal response corresponds to the possibility (33):

$$\xi \doteq \left\| \begin{matrix} 1 & 0 \\ .156 & .844 \end{matrix} \right\| \,.$$

Thus once again we have an example of the phenomenon called strong backward stability. If $\alpha = 1$ and x_1 is slightly greater than $1/2$ the y-solution is the one corresponding to the possibility (32); and that solution has $y_1 = 0$, $y_2 = 1$. The quite striking meaning of this situation in the applications was already discussed on the corresponding example of Chapter IV, Section 3: increasing the purchase of one system at the expense of the second causes the opponent to switch all his attack to the second!

This example lacks weak forward stability; we shall show in addition that the conclusion of Theorem III is false for it. This is the reason we have carried α along.

Let $\alpha \doteq 1$. The value of $D(x, y, \xi)$ corresponding to the possibility (32), neighboring $x_1 = x_2 = 1/2$, is

$$1 - e^{-h_1^L} + 1 - e^{-h_2^L} = 2 - 1.1 e^{-h_2^L} \tag{65}$$

where we have denoted by h_1^L and h_2^L the quantities (43) and (44) corresponding to the possibility (32). Since $y_1 = 0$ and $y_2 = 1$ for the possibility (32) neighboring $x_1 = x_2 = 1/2$ and for α near 1, we may rewrite h_2^L in the form

$$h_2^L = \frac{10x_1 + 100x_2 e^{-\alpha} - \log 10}{11} \tag{66}$$

for use in (65). Similarly corresponding to the possibility (33) we have

$$1 - e^{-h_1^R} + 1 - e^{-h_2^R} = 2 - 1.1 e^{-h_1^R}, \tag{67}$$

where h_1^R is given by (61) with $y = (1, 0)$ neighboring $x_1 = x_2 = 1/2$, so that we may use

$$h_1^R = \frac{10x_2 + 100x_1 e^{-1} - \log 10}{11} \tag{68}$$

in (67). The intersection satisfies

$$2 - 1.1 e^{-h_2^L} = 2 - 1.1 e^{-h_1^R} \,,$$

so that

$$h_2^L = h_1^R \,,$$

i.e., using (66) and (68),

$$x_1 + 10x_2 e^{-\alpha} = 10x_1 e^{-1} + x_2 \,. \tag{69}$$

(69) is trivially solved to yield

$$x_1 = \frac{10e^{-\alpha} - 1}{2(5e^{-1} + 5e^{-\alpha} - 1)} \qquad . \tag{70}$$

If the conclusion of Theorem III were valid, we would have

$$x_1 = \frac{\alpha}{1+\alpha}, \quad x_2 = \frac{1}{1+\alpha}, \tag{71}$$

which is not in accord with (70). If we write $\alpha = 1 + \varrho$ for ϱ near zero and write x_1 from (70) to first order terms in ϱ, we get

$$x_1 \doteq \frac{1}{2} - \frac{5e^{-1}\varrho}{20e^{-1} - 2} \doteq \frac{1}{2} - .343\varrho. \tag{72}$$

The corresponding value from (71) would be $x_1 \doteq \dfrac{1}{2} - \dfrac{\varrho}{4}$.

The question as to what general conclusions corresponding to Theorems II and III might be drawn in the presence of strong or weak backward stability is completely open.

8. *The separable problem.* In this section we return to the general problem of Section 1, with no restrictions as to the type of system, but with $\beta_{ik} = A_i c_k$. Then we get from (3)

$$b_{ik} = A_i c_k x_i' = a_i c_k \tag{72}$$

with

$$a_i = A_i x_i',$$

when the a_i is the a_i in Section 8 of Chapter VI.

In Chapter VI, Section 8, we showed that the general maximum problem there posed could in the case (72) be reduced to the determination of a γ, and in fact from the equation

$$\Sigma \frac{h_k}{c_k} = \Sigma a_i. \tag{73}$$

We saw there that the h_k were uniformly either non-increasing or decreasing in γ, with at least one h_k decreasing with respect to γ. Thus the damage function $\Sigma F_k(h_k)$ is decreasing with γ. And so is the function on the left side of (73). Thus: *the value of $\Sigma F_k(h_k)$ in Chapter VI is a monotone function of $\Sigma \dfrac{h_k}{c_k}$ and therefore of Σa_i.*

Since in the problem at hand

$$\Sigma a_i = \Sigma A_i x_i',$$

it therefore follows that the Max-Min-Max *problem reduces to the problem of finding*

$$\underset{x}{\text{Max}} \ \underset{y}{\text{Min}} \ \Sigma A_i x_i'. \tag{74}$$

Thus by the above simple argument we have reduced the Max-Min-Max problem to the linear model treated in Chapter V, and have proved the following remarkable theorem.

Theorem IV. *If* $\beta_{ik} = A_i c_k$, *the general* Max-Min-Max *problem of Section 2 decomposes into two problems;*
    (1) *that of finding*

$$\text{Max Min} \, \Sigma A_i x_i' = \gamma \tag{75}$$

subject to the constraints

$$\Sigma x_i = \Sigma y_i = 1, \quad x_i, y_i \geq 0;$$

(2) *that of solving the trivial problem of Section 8 of Chapter VI with the value of* γ *given by* (75).

See Exercise (10).

A practical remark is in order here. The typical bomb-damage formula for point targets is the survival probability

$$Q = \exp \left[-\log 2 \cdot \frac{\varrho^2}{R^2} \right]. \tag{76}$$

Here ϱ is the weapon radius and C the C.E.P. (a parameter describing the aiming error). The standard formula for ϱ is

$$\varrho = \kappa Y^{1/3}$$

where Y is the yield of the weapon and κ depends on the hardness of the target. We thus get

$$Q = \exp \left[-\frac{\log 2}{R^2} Y^{2/3} \kappa^2 \right] \tag{77}$$

or

$$Q = e^{-AC}$$

where

$$A = \frac{\log 2}{R^2} \cdot Y^{2/3}, \quad C = \kappa^2.$$

Thus the exponent separates into the product of two factors, one referring to the weapon alone and one referring to the target alone. Hence, using the method used in Section 1 of Chapter VI to fit the integer problem to the problem with real numbers, we see that in this case we are dealing with a separable problem.

Hence: for point targets, if formula (76) is admitted, we have a complete solution to the general Max-Min-Max problem, given by Theorem III; this reduces the general problem to two problems, one difficult but solved in Chapter V, the other trivial and solved in Section 8 of Chapter VI.

Exercises to Chapter VII

The author does not know the solution to the starred exercise.

1. Verify that the example of Chapter IV, Section 1, is strongly backward stable and not weakly forward stable in both admissible directions at the solution point. What about other points?

2. Verify that the example of Chapter IV, Section 2, is weakly stable both forward and backward.

3. Verify that the example of Chapter IV, Section 3, is strongly backward stable and not weakly forward stable.

4. Prove that λ_i and h_k depend continuously on x and y (Section 3).

5. In passing from the top line of the derivation (10) to the second line, we used the convention (8). Verify the legitimacy of this step.

6. Use formula (8) to prove that $\dfrac{\lambda_i(x, y)}{x_i}$ is continuous in the (x, y) space at the point $x_i = 0$.

7. Calculate $D_g H(x, y)$ [formula (13)].

8.* Attempt to find a necessary condition for $x_n > 0$ corresponding to condition (23).

9. Prove Theorem III.

10. (The separable case.) (Difficult.) Consider two weapons systems, one percentage and one numerically vulnerable. Take $\alpha_1 = \alpha_2 = 1$, so that $x_1' = x_1 e^{-y_1}$ and $x_2' = x_2 e^{-y_2/x_2}$. Take $x_1 + x_2 = 1$, $y_1 + y_2 = 1$, $x_1, x_2, y_1, y_2 \geqq 0$. Use the damage matrix

$$\|\beta_{ik}\| = \left\| \begin{matrix} 1 & 2 & 3 & 4 & 5 \\ 1.2 & 2.4 & 3.6 & 4.8 & 6 \end{matrix} \right\|$$

and suppose the targets of equal value. Reduce to a problem of the type of Chapter V, and solve that problem to determine the optimal mixture x_1, x_2 and the corresponding solution for η and ξ.

   Crib: There are only two intervals for $H(\xi)$, i.e. $\xi^* = 0$, $\xi^{**} = .708$. The value of the Max-Min, i.e. the value of $H(\xi)$ at the maximum, is .519, and γ turns out to be about 1.96. The ξ matrix is

$$\left\| \begin{matrix} 0 & .02 & .273 & .344 & .361 \\ 0 & .02 & .273 & .344 & .361 \end{matrix} \right\|.$$

Appendix

The Lagrange Multiplier Theorem
for Max-Min with several constraints

This appendix is based on a paper by J. BRAM [1].

The object of this appendix is to extend the Lagrange-multiplier result of Chapter III (Theorem VII), which was given for the simple constraint $x_1 + \cdots + x_n = X$, $x_i \geq 0$, $i = 1, \ldots, n$, to the general case of several possibly nonlinear constraints of the form

$$g_j(x) \leq 0, \quad j = 1, \ldots, m. \tag{1}$$

The result, stated as a theorem below, generalizes also the result of KUHN and TUCKER [6].

We shall maintain the hypotheses of Chapter III that the function $F(x, y)$ and the partial derivatives $F_{x_i}(x, y)$ are continuous in the set of pairs (x, y) for $y \in Y$ and for x in a fixed subset A of Euclidean space containing the set (1). In addition we assume that the $g_j(x)$ are continuously differentiable functions of x in the same set A.

Let x^0 be the point at which $\varphi(x)$ is maximized subject to the conditions (1). We use the usual notation for the gradient:

$$\nabla g_j(x) = \left(\frac{\partial g_j}{\partial x_1}, \ldots, \frac{\partial g_j}{\partial x_n} \right).$$

We denote by J the set of j for which $g_j(x^0) = 0$.

The direction γ is said to be *possible* at x^0 if there is an arc with that direction issuing from x^0 with direction γ in the sense of the Remark following Theorem I of Chapter III, lying entirely in A. The direction γ is *admissible* at x^0 if it is possible and there is an arc isuing from x^0 lying entirely in the constraint space defined by (1).

We denote by Γ the set of directions admissible at x^0. The directional derivative at x^0 exists along these arcs, and does not depend on the particular choice of arc having a direction $\gamma \in \Gamma$, so that

$$D_\gamma \varphi(x) \leq 0 \quad \text{for} \quad \gamma \in \Gamma. \tag{2}$$

Since these arcs lie in the space (1), then also, if $\gamma \in \Gamma$

$$\nabla g_j(x^0) \cdot \gamma \leq 0 \quad \text{for} \quad j \in J. \tag{3}$$

We denote the set of all possible γ satisfying (3) by $\varDelta$. We have just observed that $\varGamma \subset \varDelta$.

We now impose the Kuhn-Tucker constraint qualification, i.e. we assume that $\varGamma = \varDelta$. This means that to every possible direction γ satisfying (3) there corresponds an arc issuing from x^0 in that direction and lying in the constrained set (1). This is obviously true in the case when the g_j and A are linear. What it does in the general case is to eliminate pathologies such as certain types of cusps.

Thus with the Kuhn-Tucker constraint qualification assumed we may restate (2) as follows:

$$D_\gamma \varphi(x) \leqq 0 \quad \text{for all} \quad \gamma \in \varDelta. \tag{4}$$

Let now W be the set of vectors

$$\left(F_{x_1}(x^0, y), \ldots, F_{x_n}(x^0, y) \right)$$

for $y \in Y(x^0)$. Formula (12) of Chapter III may be rewritten

$$D_\gamma \varphi(x) = \operatorname*{Min}_{w \in W} \gamma \cdot w, \tag{5}$$

the minimum being achieved.

Let now $\hat{W}$ be the convex hull of W, i.e., the set of vectors $\hat{w}$ representable in the form

$$\hat{w} = \sum_{k=1}^{s} \alpha_k w_k$$

with $\Sigma \alpha_k = 1$, $\alpha_k \geqq 0$, $w_k \in W$, $k = 1, \ldots, s$. Obviously

$$D_\gamma \varphi(x) = \operatorname*{Min}_{\hat{w} \in \hat{W}} \gamma \cdot \hat{w}. \tag{6}$$

Now define the set $\varDelta^*$ dual to $\varDelta$ by the condition that $w^* \in \varDelta^*$ if and only if

$$w^* \cdot \gamma \leqq 0 \quad \text{for all} \quad \gamma \in \varDelta. \tag{7}$$

We are now ready to prove the following theorem.

Theorem. *If x^0 maximizes $\varphi(x)$ subject to the constraints* (1), *and if the Kuhn-Tucker constraint qualification is satisfied, then there exist $\lambda_1, \ldots, \lambda_m$, all non-negative, such that*

$$D_\gamma \varphi(x) \leqq \sum_{j=1}^{m} \lambda_j \gamma \cdot \nabla g_j(x^0)$$

for all possible γ. Further, if $g_j(x^0) < 0$, $\lambda_j = 0$.

Proof. We begin by proving that $\varDelta^* \cap \hat{W}$ is not empty. If this intersection were empty, there would be a shortest segment of positive length joining the compact set $\hat{W}$ to the closed set $\varDelta^*$. Let the equation of the perpendicular bisector $\varPi$ of this segment be

$$\gamma^0 \cdot z = \sum_{i=1}^{n} \gamma_i^0 z_i = c. \tag{8}$$

Evidently neither $\varDelta^*$ nor $\hat{W}$ meets $\varPi$: we may therefore suppose that $\gamma^0 \cdot z < c$ on $\varDelta^*$ and $\gamma^0 \cdot z > c$ on $\hat{W}$. Since $0 \in \varDelta^*$, $0 < c$.

Let now $z \in \varDelta^*$. Suppose $\varepsilon > 0$. Then $\left(\dfrac{c}{\varepsilon}\right) z \in \varDelta^*$, so that $\gamma^0 \cdot \left(\dfrac{z}{\varepsilon}\right) < c$, i.e., $\gamma^0 \cdot z < \varepsilon$. It follows that for any $z \in \varDelta^*$, $\gamma^0 \cdot z \leq 0$. Now in particular $\nabla g_j(x^0) \in \varDelta^*$, by the definition (7), for all $j \in J$. Hence

$$\gamma^0 \cdot \nabla g_j(x^0) \leq 0 \quad \text{for all} \quad j \in J. \tag{9}$$

On referring to the definition of $\varDelta$ [(3) and the first sentence following], we see that (9) implies that $\gamma^0 \in \varDelta$. Thus, from (2), $D_{\gamma^0} \varphi(x^0) \leq 0$, so that, using (5), we have a $w \in W$ such that $\gamma^0 \cdot w \leq 0$. But this contradicts the statement above that $\gamma^0 \cdot z > c > 0$ for all $z \in \hat{W}$. Hence $\varDelta^* \cap \hat{W} \neq \varLambda$.

Now let $z \in \varDelta^* \cap \hat{W}$. Since $z \in \varDelta^*$, (7) holds with $w^* = z$. Since $\varDelta$ is defined by the finite set of linear inequalities (3), we may apply the theorem of FARKAS [2][14]; there exist non-negative λ_j corresponding to $j \in J$ such that $z = \sum_{j \in J} \lambda_j \nabla g_j(x^0)$. By putting $\lambda_j = 0$ for $j \notin J$, we may write this as follows:

$$z = \sum_{j=1}^{m} \lambda_j \nabla g_j(x^0).$$

Now we are ready to use the fact that $z \in \hat{W}$. This fact combined with (6) yields, for any possible γ,

$$D_\gamma \varphi(x) = \operatorname*{Min}_{\hat{w} \in \hat{W}} \gamma \cdot w \leq \gamma \cdot z. \tag{10}$$

But (10) is exactly the statement of the theorem.

---

[14] This theorem is the following: Write $a = (a_1, \ldots, a_n)$, $u = (u_1, \ldots, u_n)$ and $a \cdot u = \Sigma a_i u_i$. Given a collection $a^1, \ldots, a^p$, let S be the set of u such that

$$a^1 \cdot u \geq 0$$
$$\vdots$$
$$a^p \cdot u \geq 0.$$

Suppose that $b \cdot w \geq 0$ for all $u \in S$. Then there exists a set $\lambda_1, \ldots, \lambda_p$ of non-negative numbers such that

$$b = \sum_{k=1}^{p} \lambda_k a^k.$$

Bibliography

[1] BRAM, J.: The Lagrange multiplier theorem for max-min with several constraints, SIAM Journal, vol. 14, pp. 665—667 (1966).
[2] FARKAS: Über die Theorie der einfachen Ungleichungen. J. reine angew. Math. **124**, 1—27 (1901).
[3] GRAVES, L. G.: The theory of functions of real variables. New York: McGraw-Hill (1946).
[4] KAKUTANI, S.: A generalization of Brouwer's fixed point theorem. Duke Math. J. **8**, No. 3, 457—459 (1941).
[5] KOOPMAN, B. O.: The theory of Search III, optimum distribution of searching effort. Operations Res. **5**, 613—626 (1957).
[6] KUHN, H. W., and A. W. TUCKER: Nonlinear programming. Second Berkeley Symposium on Mathematical Statistics and Probability. University of California Press (1951).
[7] HARLAN MILLS: Marginal values of matrix games and linear programs. Ann. Math., Study No. 38. Linear inequalities and related systems. 183—193 (1956).
[8] DANSKIN, JOHN M.: The theory of max-min, with applications. SIAM Journal, vol. 14, pp. 641—664 (1966).

Fotosatz, Druck und Bindearbeit: Brühlsche Universitätsdruckerei Gießen

SPRINGER-VERLAG
BERLIN · HEIDELBERG · NEW YORK

Ökonometrie und Unternehmensforschung
Econometrics and Operations Research

Herausgegeben von/Edited by

M. BECKMANN, Bonn; R. HENN, Göttingen; A. JAEGER, Cincinnati; W.
KRELLE, Bonn; H. P. KÜNZI, Zürich; K. WENKE, Ludwigshafen; PH.
WOLFE, Santa Monica (Cal.)
Geschäftsführende Herausgeber/Managing Editors
W. KRELLE und H. P. KÜNZI